FLYING IN CONGESTED AIRSPACE

Other TAB Books by the author:

No. 2301 *Upgrading Your Airplane's Avionics*
No. 2308 *How to Become an Airline Pilot*

FLYING IN CONGESTED AIRSPACE

BY TIMOTHY R. V. FOSTER

To Bozo Two

FIRST EDITION

FIRST PRINTING

Printed in the United States of America

Library of Congress Cataloging in Publication Data

Foster, Timothy R. V.
Flying in congested airspace.

Includes index.
1. Airplanes—Piloting. 2. Airports—Traffic control.
3. Airplanes—Collision avoidance. I. Title.
TL710.F618 1983 629.132'5214 83-4913
ISBN 0-8306-2358-2 (pbk.)

Charts in Appendix B reproduced with permission of Jeppesen Sanderson, Inc.

Cover photograph courtesy of Aircraft Radio and Control Division, Cessna Aircraft.

Contents

Introduction

Flying in congested airspace is different from flying elsewhere. Consider these questions:

You must be better trained and able to handle complex demands decisively and without faltering. What are the requirements, and how do you ''practice up?''

Your airplane must be specially equipped for many operations. What equipment is required? How do you operate if some of the special equipment fails or is unavailable?

There are restrictions as to where you may go. How do you learn about these so you can comply with the regulations?

Major airline airports, such as Kennedy, O'Hare, Washington National, Atlanta, Los Angeles International and others all have special procedures unique to their own locations. How do you find out about them?

What is the correct communications process for arriving and departing at a major airline airport? How do you track the radio frequencies you'll need?

What jurisdictions must you deal with in high density areas? How do you reach them by radio? How do you arrange an orientation visit to ATC?

What do you do when you see you're going to pass behind a 747 a mile away and at your altitude?

What are the safest ways to avoid wake turbulence on takeoff? On landing? What must you watch out for when taxiing?

Tower clears you for immediate takeoff 30 seconds behind a 727 that just left the same runway. There are five airliners lined up behind you and a DC 10 on final. What do you do?

What are the military low-level training routes, and how do they operate? How do they affect you? What can you do to maintain maximum safety?

How do you avoid delays, while staying legal and safe?

How do you go where you want to, in spite of restrictions?

How do you avoid devious routings, saving time, fuel, and money?

How do you drop off/pick up passengers at airline airports?

Landing or taking off at Kennedy in a Cessna 172 can cost you $52 or $20, depending on when you do it. What's best?

Which peripheral airports are best in each area to avoid high landing fees, air traffic delays, high ground transportation costs? Which ones are open all night?

Each chapter in this book except the last ends with a brief quiz. Why not try taking the quizzes first? Then you'll know what you need to know.

Chapter 1
What Is Congested Airspace?

Congested airspace is where you've really got to have your wits about you when you're flying. This book will show you how to fly safely and efficiently in the busiest areas for air traffic in the United States.

These areas are defined as any *Terminal Control Areas (TCAs), Terminal Radar Service Areas (TRSAs),* major airline and busy general aviation airports, areas supporting special events (e.g., the annual Oshkosh fly-in in August, or wherever the Super Bowl is), and military training areas.

HOW MUCH CONGESTED AIRSPACE IS THERE?

In the spring of 1982, there were 23 Terminal Control Areas in the U.S.A., 66 Terminal Radar Services Areas, and 470 Military Training Routes. There were some 500 airports with control towers, and over 100 airports with more than 20,000 movements a year. Due to traffic congestion, flight within 30 control zones in the U.S. is not allowed under Special Visual Flight Rules (SVFR), which permit operations under lower-than-VFR visibility and cloud clearance minima.

The United States consists of over 3.6 million square miles. There are over 210,000 airplanes registered in the U.S., which means that there are over 17 square miles for

American Airlines Boeing 727 frames La Guardia's unique Control Tower. (courtesy Port Authority of NY & NJ)

each airplane at any given altitude. And, of course, *all* the airplanes are never in the air at the same time. Nevertheless, each year there are about 30 midair collisions, and nobody knows how many near-misses.

Congested airspace is the result of heavy air traffic. The heaviest areas are those where TCAs are to be found.

There are presently two types of TCA—Group I and Group II. A third type, Group III, is provided for in FAR 91.90, but at present, none of these exists. The Group designation determines the equipment and operational requirements of aircraft that may fly in them. As of 1982, these were the TCAs in the U.S.:

Terminal Control Areas—Group I (see Appendix A)

Atlanta, Georgia
Boston, Massachusetts
Chicago, Illinois
Dallas/Fort Worth, Texas
Los Angeles, California
Miami, Florida
New York, New York
San Francisco, California
Washington, D.C. (Virginia)

Terminal Control Areas—Group II

Cleveland, Ohio
Denver, Colorado
Detroit, Michigan
Honolulu, Hawaii
Houston, Texas
Kansas City, Missouri
Las Vegas, Nevada
Minneapolis, Minnesota
New Orleans, Louisiana
Philadelphia, Pennsylvania
Pittsburgh, Pennsylvania
St. Louis, Missouri
San Diego, California
Seattle, Washington

REQUIREMENTS FOR FLIGHT WITHIN TCA

Regardless of weather conditions, you must have an Air Traffic Control (ATC) clearance prior to entering a TCA. You may not even *request* such a clearance unless you meet the requirements of Federal Aviation Regulations (FAR) 91.24 and 91.90. Included among these requirements are:

Beech Musketeer inside the St. Louis TCA. (courtesy Beech)

Group I TCAs

- ☐ A two-way radio capable of communicating with ATC on appropriate frequencies.
- ☐ A VOR or TACAN receiver, except for helicopters.
- ☐ A 4096 code transponder with Mode C automatic altitude reporting equipment, except for helicopters operating at or below 1,000 feet above ground level (AGL) under a Letter of Agreement. [Note: It *is* possible to obtain a deviation from these requirements by making a request to ATC. A lack of altitude reporting equipment may be permitted by a simple radio request on the TCA frequency, but flight without transponder must be requested at least four hours before the proposed operation.]
- ☐ A Private Pilot Certificate or better if you want to land or take off from an airport within the TCA.
- ☐ A speed limit of 200 knots.

Group II TCAs

- ☐ As above, except:

- ☐ The transponder need not have Mode C altitude reporting capability.
- ☐ You may land at an airport within the TCA with as little as a Student Pilot Certificate.

Group III TCAs

- ☐ If two-way radio communications are maintained with ATC while within the TCA and you provide position, altitude and proposed flight path prior to entry, you do not need a transponder or VOR/TACAN receiver. (As of 1982, none exists.)

AIRSPACE USERS' NEEDS

To understand how TCAs work, you have to have the whole picture. Airspace users' operations and needs are varied:

Airliners and executive jets want to get up to high altitudes (25,000 to 45,000 feet) quickly, stay as high as possible until close to the destination, and then descend

Gulfstream III business jet. (courtesy Gulfstream-American)

quickly, with no delays, to the airport. Some executive jets can cruise at 50,000 feet. Cruising speeds—Mach .6 to Mach .8 (450 to 550 knots). Approach speeds—120 to 150 knots.

SSTs (i.e., Concorde) are prohibited from supersonic flight over land in the U.S., but make their transition to and from flight at Mach 2 just offshore, at altitudes ranging from 29,000 to 60,000 feet. At the moment, they are only operating out of New York and Washington, and the offshore areas east and north of those airports. Cruising speed—Mach 2 (1,200 knots) to Mach .8 (about 550 knots) at lower altitudes. Approach speeds—160 to 190 knots.

Commuter airliners need direct routings between high and low density traffic areas, at altitudes ranging up to 18,000 feet or more, but many flight stay below 12,000 feet. Cruising speeds—180 to 300 knots. Approach speeds—100 to 120 knots.

Turboprop executive aircraft want cruising altitudes between 12,000 and 24,000 feet. Cruising speeds

Concorde—a view to avoid. (courtesy British Airways)

Beech Queen Air commuter airliner. (courtesy Beech)

—180 to 300 knots. Approach speeds—100 to 120 knots.

Piston-powered general aviation aircraft want the airspace below 10,000 feet (turbocharged aircraft can operate in the same airspace as turboprops—up to about 24,000 feet). The first 5,000 feet or so are needed for local and training flights, while the remainder are needed for

Piper Cheyenne III corporate turboprop at San Jose. (courtesy Piper)

Beech 58 Baron executive twin. (courtesy Beech)

cross-country flying. Cruising speeds—100 to 200 knots. Approach speeds—60 to 100 knots.

Military aircraft need very high altitudes (45,000 feet and up) and large chunks of airspace at all altitudes where they can practice aerobatics, bombing or strafing runs, in-flight refueling, and low-level cross-country mis-

Boeing KC135 refueling a brace of F16s. (courtesy USAF)

sions in all weather. Cruising speeds—anywhere from 200 knots up to Mach 2 and beyond. Approach speeds—anywhere from 100 to 200 knots.

Helicopters want low altitudes—up to about 5,000 feet. They generally operate at speeds between 80 and 150 knots.

Gliders and sailplanes generally operate below 5,000 feet, but some high-performance sailplanes go up to the low 20s or more—mostly in mountainous areas.

Hot-air balloons and hang gliders generally fly below 1,000 feet and in conditions of low wind.

That's quite a mix! It's not bad when everybody is segregated. It's when they all come together that problems arise. Thus, where mixes are likely to occur, restrictions must be placed on operations for safety reasons. The complexity or density of aircraft movements in some airspace areas result in special aircraft and pilot requirements for operation within that airspace. Pilots must be familiar with the operational requirements for the various airspace segments.

Hughes 500D in downtown Los Angeles. (courtesy Hughes)

Schweizer I-26 sailplane. (courtesy Schweizer)

UNCONTROLLED AIRSPACE

Uncontrolled airspace is that portion of the airspace in which ATC provides no separation or clearance of air traffic. There is not much of it left in the U.S.—mostly airspace below 14,500 feet (virtually all airspace above 14,500 feet is controlled) in which there is no control area,

Hot-air balloons by Raven. (courtesy Raven)

airway, control zone, terminal control area, or transition area. The rule of traffic avoidance in uncontrolled airspace is *see and be seen*. Flight under the instrument flight rules (IFR) is permitted in uncontrolled airspace, but there is no assurance of traffic separation. Pilots must be instrument rated to fly IFR, even in uncontrolled airspace.

Although traffic is light in uncontrolled airspace, this is no reason to relax vigilance when flying. Keep your eyes open for other traffic at all times. No small number of midair collisions have occurred in uncontrolled airspace over the years. Pilots tend to become lax as far as adherance to cruising altitudes is concerned. Don't be one of them. Always be especially watchful as you fly over a navigational aid, such as a VOR, since the convergence of routes over a VOR tends to collect aircraft. Also, VORs are often where people practicing instrument flying do a lot of air work, such as holding. Given that they are probably paying less attention to the outside world, make up for it by paying *more* attention yourself.

CONTROLLED AIRSPACE

Controlled airspace is that portion of the airspace in which ATC provides separation or clearance of air traffic. It consists of Control Areas, Airways, Control Zones, Terminal Control Areas, or Transition Areas. The rules of traffic avoidance in controlled airspace are *see and be seen* for VFR traffic and IFR aircraft operating in visual meteorological conditions (VMC). IFR aircraft are always given separation from *other* IFR aircraft in VMC *or* IMC, and from VFR aircraft in positive control areas.

The rules for VFR weather minima are more stringent in controlled airspace than in uncontrolled airspace (Tables 1-1 and 1-2).

Control Zones

Control Zones are designed to provide controlled airspace around one or more airports where traffic density warrants. There is usually (but not always) a control tower at an airport with a Control Zone, and there is an instrument approach procedure established for the airport. They ex-

Table 1-1. VFR in Uncontrolled Airspace—Minimum Visibility and Distance from Clouds.

Flight Altitude	Flight Visibility	Distance From Clouds
1,200′ or less above the surface, regardless of MSL altitude	1 statute miles*	Clear of clouds
More than 1,200′ above the surface, but less than 10,000′ MSL	1 statute mile	500′ below 1,000′ above 2,000′ horizontal
More than 1,200′ above the surface and at or above 10,000′ MSL	5 statue miles	1,000′ below 1,000′ above 1 statute mile horizontal

*Helicopters may operate with less than 1 mile visibility, outside controlled airspace at 1,200 feet or less above the surface, if operated at a speed that allows adequate opportunity to see any air traffic or obstructions in time to avoid collisions.

Table 1-2. VFR in Controlled Airspace—Minimum Visibility and Distance from Clouds.

Flight Altitude	Flight Visibility**	Distance From Clouds**
1,200′ or less above the surface, regardless of MSL Altitude	3 statute miles	500′ below 1,000′ above 2,000′ horizontal
More than 1,200′ above the surface, but less than 10,000′ MSL	3 statute miles	500′ below 1,000′ above 2,000′ horizontal
More than 1,200′ above the surface, and at or above 10,000 MSL	5 statute miles	1,000′ below 1,000′ above 1 statute mile horizontal

**When operating within a Control Zone beneath a ceiling, the ceiling must be at least 1,000′. If the pilot wants to land or take off or enter a traffic pattern within a Control Zone, the *ground* visibility must be at least 3 miles at that airport. If ground visibility is not reported at the airport, 3 miles *flight* visibility is required. (FAR 91.105) tend upward from the surface of the Earth, terminating at the base of the Continental Control Area (14,500 feet). A few Control Zones do not underlie the Continental Control Area, and thus have no upper limit. A Control Zone may include one or more airports and is normally a circular area with a radius of five statute miles and any extensions necessary to include instrument departure and arrival paths.

Control Zones are depicted on charts and if effective only during certain hours of the day, this will be noted on the chart.

Special VFR

In some Control Zones it is possible to operate an aircraft in VFR flight below the VFR weather minima outlined below for controlled airspace. This is called *Special VFR (SVFR)*. The purpose of SVFR is to allow you to get out of or get into an airport when the weather is marginal, but not solid IFR. In effect, it enables you to get to or from uncontrolled airspace (perhaps five statute miles away) legally, where the VFR minima are more liberal. Special VFR requires a clearance from ATC. You won't get an SVFR clearance if there is any IFR traffic operating. However, I've often seen situations where there is early morning fog and a lineup of IFR aircraft awaiting departure clearance, and an SVFR type squeaks out with no waiting, because there is no IFR traffic in the air at the time.

Special VFR Weather Minima

- ☐ All flight must be clear of clouds.
- ☐ Flight visibility must be at least one statute mile (except for helicopters).
- ☐ Ground visibility at the airport is at least one statute mile (flight visibility may be used if there is no way of obtaining the ground visibility).
- ☐ If at night, the aircraft and pilot must be equipped and legal for IFR flight.

Airports Where SVFR Is Not Authorized

Thirty airports are so busy that you can't get an SVFR

clearance within their Control Zones at all. They are:

ATL—Atlanta, GA (The Hartsfield International Airport)
BAL—Baltimore, MD (Baltimore-Washington International Airport)
BOS—Boston, MA (Logan International Airport)
BUF—Buffalo, NY (Greater Buffalo International Airport)
ORD—Chicago, IL (O'Hare International Airport)
CVG—Cincinnati, OH (Greater Cincinnati Airport)
CLE—Cleveland, OH (Cleveland-Hopkins International Airport)
CMH—Columbus, OH (Port Columbus International Airport)
DFW—Dallas, TX (Dallas-Ft. Worth Regional Airport)
DAL—Dallas, TX (Love Field)
DEN—Denver, CO (Stapleton International Airport)
DTW—Detroit, MI (Detroit-Metro Wayne County Airport)
IAH—Houston, TX (Houston Intercontinental Airport)
IND—Indianapolis, IN (Indianapolis Municipal Airport)
LAX—Los Angeles, CA (Los Angeles International Airport)
SDF—Louisville, KY (Standiford Field)
MEM—Memphis, TN (Memphis International Airport)
MIA—Miami, FL (Miami International Airport)
MSP—Minneapolis, MN (Minneapolis-St. Paul International Airport)
EWR—Newark, NJ (Newark International Airport)
JFK—New York, NY (Kennedy International Airport)
LGA—New York, NY (La Guardia Airport)
MSY—New Orleans, LA (New Orleans International Airport)
PHL—Philadelphia, PA (Philadelphia International Airport)
PIT—Pittsburgh, PA (Greater Pittsburgh Airport)

PDX—Portland, OR (Portland International Airport)
SFO—San Francisco, CA (San Francisco International Airport)
SEA—Seattle, WA (Seattle-Tacoma International Airport)
STL—St. Louis, MO (Lambert-St. Louis International Airport)
TPA—Tampa, FL (Tampa International Airport)
DCA—Washington, DC (Washington National Airport)

Transition Areas, Control Areas, Airways

These Control Areas consist of the airspace designated as Colored Federal Airways (e.g., Red 1, Amber 2, based on low/medium frequency navigational aids), VOR Federal Airways (e.g., Victor 282, Victor 5, etc.), and Control Area Extensions, but do not include the Continental Control Area. Unless otherwise designated, Control Areas also include the airspace between a segment of a main VOR Airway and its associated alternate segments. Airways generally start at 1,200 feet or more above the ground, but some transition areas start at 700 feet above the ground. They top out at the base of the Continental Control Area.

Continental Control Area

The Continental Control Area consists of the airspace of the 48 contiguous States, the District of Columbia and most of Alaska, at and above 14,500 feet MSL, but does not include: the airspace less than 1,500 feet above the surface of the Earth, or most Prohibited and Restricted Areas.

In other words, virtually *all* airspace over the U.S. above 14,500 feet is controlled. VFR flight without a clearance is permitted within the Continental Control Area in VMC.

Positive Control Area

The Positive Control Area is basically all airspace over the U.S. between 18,000 feet and flight level 600

(60,000 feet). A flight level (FL) is the altitude read by an altimeter, expressed in hundreds of feet when it is set to standard pressure of 29.92 inches rather than the local altimeter setting for the region. All flight above 18,000 feet is so operated, and all altitudes above 18,000 feet are called flight levels. For example 20,000 feet is referred to as FL 200. 43,500 feet is FL 435. VFR flight within the Positive Control Area is not permitted at any time. All flight must be in accordance with an IFR clearance.

An additional requirement is that all aircraft operating above FL 240 must be equipped with Distance Measuring Equipment (DME) when VOR equipment is required.

SPECIAL USE AIRSPACE

Special Use Airspace is the airspace where either flight operations are confined because of their nature (e.g., test flying areas), where limitations are imposed upon aircraft operations that aren't part of those activities, or both. Except for controlled firing areas, Special Use Airspace areas are shown on aviation charts. Special Use Airspace includes:

- ☐ Airport Traffic Areas
- ☐ Prohibited Areas
- ☐ Restricted Areas
- ☐ Warning Areas
- ☐ Military Operations Areas
- ☐ Alert Areas
- ☐ Controlled Firing Areas
- ☐ Military Training Routes
- ☐ Temporary Flight-Restriction Areas

Airport Traffic Areas

Airport Traffic Areas consist of the airspace within a horizontal radius of five statute miles from a geographical center of any airport at which a control tower is operating, extending from the surface up to, but not including, an altitude of 3,000 feet above the elevation of the airport, unless otherwise specifically designated in FAR Part 93.

Airport Traffic Areas *are not* depicted on charts.

If you are not landing or taking off from an airport within the airport traffic area, you must avoid the area unless you have authorization from ATC. This may be obtained by contacting the appropriate tower, or from a facility from which you are receiving radar services, such as an approach control unit.

If you are operating to or from an airport served by a control tower, you must establish and maintain radio communications with the tower prior to and while operating in the airport traffic area. And you must maintain radio communications with the tower while operating on the movement areas of the airport.

Maximum indicated airspeeds within Airport Traffic Areas are prescribed by FAR 91.70. Piston engine aircraft may not exceed 156 knots (180 mph) indicated airspeed (IAS), while turbine-powered aircraft may not exceed 200 knots (230 mph). These speed limits don't apply within a Terminal Control Area, where the speed limit, below 10,000 feet, is 250 knots (288 mph) IAS.

Prohibited Areas

Prohibited areas are chunks of airspace identified by an area on the surface of the Earth within which the flight of aircraft is prohibited. For example, in Washington DC, flight is prohibited over the White House and the Mall area. These Prohibited Areas are established for security or other reasons associated with the national welfare. They are shown on aviation charts.

Restricted Areas

Flight within Restricted Areas is not totally prohibited, but it *is* subject to restrictions. The reason is the kind of activity going on in the area. For example, missile firing ranges are located in Restricted Areas, as is the Kennedy Space Center at Cape Canaveral, Florida. Flying within Restricted Areas without authorization from the using or controlling agency may be extremely hazardous to your health! Some Restricted Areas have limitations denoted by time or day of the week, e.g., "Active 0700-2300 LT (local time)daily;""Intermittent VFR weekdays only."

Why they have Restricted Areas—the space shuttle. (courtesy NASA)

Sometimes the controlling agency is shown beside the Restricted Area legend: "0500-2300 LT Mon-Fri, 1,200/ GND JAX TRACON," which means that the area is restricted between 5 a.m. and 11 p.m., Mondays to Fridays, between 1,200 feet MSL and the ground, under the jurisdiction of the Jacksonville Terminal Radar Control Unit.

Warning Areas

Warning Areas are basically the same thing as Restricted Areas, but they are found only in *international* airspace. They are established beyond the three-mile limit. Though the activities conducted within warning areas may be hazardous as those in Restricted Areas, Warning Areas cannot be legally designated because they are over international waters. Hours of operation and controlling agencies are depicted on charts in a similar manner to those for Restricted Areas.

Military Operations Areas (MOAs)

Military Operations Areas consist of airspace of defined vertical and lateral limits established for the purpose of separating certain military training activities from IFR traffic. Whenever a MOA is being used, nonparticipating IFR traffic may be cleared through if IFR separation can be provided by ATC. Otherwise, ATC will reroute or restrict nonparticipating IFR traffic.

Some training activities may necessitate acrobatic maneuvers, and the military is exempted from the regulation prohibiting acrobatic flight on airways within MOAs.

When operating VFR, you should exercise extreme caution while flying within a MOA when military activity is being conducted. Information regarding activity in MOAs may be obtained from any Flight Service Station (FSS) within 100 miles of the area. Most MOAs are shown on aviation charts.

Alert Areas

Alert areas are depicted on aviation charts to inform nonparticipating pilots of areas that may contain a high volume of pilot training or an unusual type of aerial activity. You should be particularly alert when flying in these areas. All activity within an Alert Area is conducted in accordance with Federal Aviation Regulations, without waiver, and pilots of participating aircraft as well as pilots transiting the area are equally responsible for collision avoidance.

MOA activity—a quartet of F-15s. (courtesy USAF)

Aerobatic Eagle by Christen—something to watch out for in an Alert Area. (courtesy Christen).

Controlled Firing Areas

Controlled Firing Areas contain activities which, if not conducted in a controlled environment, could be hazardous to nonparticipating aircraft. A Controlled Firing Area suspends its activities immediately when spotter aircraft, radar, or ground lookout positions indicate an aircraft might be approaching the area. Controlled Firing Areas are not charted since they do not cause a nonparticipating aircraft to change its flight path.

Military Training Routes (MTRs)

(Reprinted from *Avemco Pilot Bulletin* June 1982, by permission.)

Last seen in a Controlled Firing Area—Beech AQM-37 missile target. (courtesy Beech)

One of the best ways to scare the living daylights out of an unsuspecting pilot is for an Air Force F-111 or F-4 to appear out of nowhere and flash near your aircraft at 250 knots or better. And yet, there's a very good chance that

MTR traffic—an A-10 Thunderbolt II. (courtesy USAF)

hot military jet fighters are out there somewhere, even though many general aviation pilots may never see them. Aircraft camouflaging is often masterfully rendered. Like the survival-minded chameleon, military fighters blend cleverly against drab background environs, there to ferret out and vanquish their prey. The idea, of course, is to make sure you're not the accidental prey. And the first safeguard is to be mindful of the existence of military training routes (MTRs) that cross or come near your intended route of flight.

In recent years, the FAA has recorded an estimated 100 military-civilian "near-misses" a year. There also have been a number of fatal midairs. It is possible, too, that some civilian aircraft downed for unexplained reasons could have survived a near-miss only to be sent out of control or broken up as a result of tornadic wingtip vortices generated by 600-mph military aircraft.

In terms of ensuring pilot preparedness, training and special mission flights over these routes are essential to the nation's defense. To maintain sharp pilot proficiency, pilots of various military aircraft must practice, in the real environment, missions such as aircraft intercept, air-to-air combat and photo reconnaissance. Routes frequently are flown close to the ground to simulate penetration of enemy radar. Oftentimes, then, military aircraft barrel along at near supersonic speed through low-altitude, often unrestricted airspace shared by general aviation pilots. This can be a distressing fact of aviation life, especially considering that many civilian pilots aren't even aware MTRs exist. Many other pilots who know of the MTR routes do not give them proper respect during preflight.

There are two basic types of military training routes: VRs, or VFR routes, and IRs, IFR-only routes. Most VRs are flown from the surface to 1,500 feet above ground level, with other segments of the same route calling for altitudes of, perhaps, up to 10,000 feet. Likewise, route widths may vary from two or three miles to 15 miles.

IRs, on the other hand, are generally flown at altitudes of between 2,000 and 7,000 feet MSL. Most of the time, IR flights are conducted under ATC clearance.

How is the general aviation pilot to know where these routes lie? You can obtain a copy of a special chart called the "Green Demon" that displays all the VR and IR routes in service. The Demon consists of a set of three charts covering the entire U.S. It is issued every 56 days, along with special flight information publication (FLIP) AP/1B.

In early 1979, the FAA began displaying VRs and IRs on its low-altitude en route charts, except those VRs at or below 1,500 feet AGL. They are depicted in light brown, with the route number and an arrow showing the direction of flight. A legend also informs pilots of the altitude ranges at which each of the routes may be flown.

Lately, sectional charts have started depicting MTRs. At last, VFR pilots have ready access to MTR information. The routes are shown in light gray, are numbered by type (e.g., VR 714) and also show the direction of flight with an arrow. Unlike the routes shown on the low-altitude charts, however, sectional charts do not provide MTR altitude information. Also be aware that while sectional charts are updated every six months, MTRs are updated every 56 days; chart information could therefore be out of date.

Knowing the route number, ask the FSS briefer if the route will be in use during the time of the intended flight. The briefer will call up the route information by computer and quickly tell whether the MTR will be "hot," or in use, and what the affected altitudes and times are. Some words of caution, though. Briefers won't tell you about MTRs unless you ask, and also, charts show route centerlines only. And lastly, it is possible for a route to be inactive. So it's best to check route status enroute, too. While the routes don't take long to fly, they're often in use more than once during weekdays and Saturdays (rarely on Sundays). By reviewing the in-use time with the briefer, pilots will be able to select a route, altitude and flight time that will steer them clear of military traffic.

Once airborne and flying in the vicinity of an MTR, it's still a good idea to maintain a heads-up vigil against any 600-mph surprises. Many military aircraft literally zip around like bullets in the air. In fact, fighters capable of 900 feet-per-second speeds are moving as fast as a .45-caliber slug the moment it leaves the gun barrel.

Should you spot one aircraft, be on the lookout for others. Often, missions are flown with, perhaps, as many as four aircraft. Depending again on the maneuvers called for in the mission, the flight paths can vary from straight-and-level to aerobatic-type maneuvers. You can be somewhat assured that military pilots should be looking out for you, too. Or at least that's the flight plan. Fighter pilots flying MTRs are specifically briefed to spend about 95 percent of their time watching out for checkpoints and other visual reference. (End of Avemco reprint.)

Military training routes are identified as follows:

- ☐ Instrument Routes (IRs) and Visual Routes (VRs) at or below 1,500 feet AGL (with no segment above 1,500) are identified by a four-digit number, e.g., IR 1007, VR 1352.
- ☐ IRs/VRs above 1,500 feet AGL (segments of these routes can be below 1,500) are identified by three digit numbers, e.g., IR 009, VR 029.

TEMPORARY FLIGHT RESTRICTIONS

Another factor affecting your flying operations are the temporary flight restrictions that may be put into effect from time to time. These may occur in the vicinity of an incident or event which might generate a high degree of public interest. For example, on July 4, 1976, the Tall ships, a huge fleet of sailing vessels, visited New York Harbor in celebration of America's bicentennial. Flight by nonessential aircraft in the airspace over the harbor was restricted by a Notice to Airmen (NOTAM). These restrictions may also be implemented because of disasters or other special events, such as demonstrations, riots, and civil disturbances, as well as major sporting events, parades, pageants, and similar functions which are likely to attract large crowds and thus airborne spectators.

The FAA declares a temporary flight restriction by NOTAM, and this contains a description of the area in which the restrictions apply. Normally the area will include the airspace below 2,000 feet above the surface within five miles of the site of the incident. However, the exact dimensions will be included in the NOTAM.

You are not allowed to operate within this type of area unless you are flying one of the following:

- ☐ An aircraft participating in disaster relief, being operated under the direction of the agency responsible for relief activities.
- ☐ An aircraft operating to or from an airport within the area which will not hamper or endanger relief activities.
- ☐ An aircraft operating under an ATC IFR clearance.
- ☐ An aircraft unable to fly around the area because of weather or other considerations; you give advance notice to the ATC facility specified in the NOTAM, and enroute flight through the area will not hamper or endanger relief activities.
- ☐ An aircraft carrying accredited news representatives or persons on official business concerning the incident, and the flight is conducted in accordance with FAR 91.79 and a flight plan is filed with the ATC facility specified in the NOTAM.

AIRSPACE SUMMARY

If you didn't realize it before, perhaps now you have a good grasp of the complex array of airspace that exists within the United States. Here is a summary of U.S. airspace:

Controlled Airspace

- ☐ Control Zones
- ☐ Terminal Control Areas
  - —Group I
  - —Group II
- ☐ Airport Traffic Areas
- ☐ Transition Areas
- ☐ Airways
- ☐ Continental Control Area

☐ Positive Control Area

Special Use Airspace

☐ Prohibited Areas
☐ Restricted Areas
☐ Warning Areas
☐ Military Operations Areas
☐ Alert Areas
☐ Controlled Firing Areas
☐ Military Training Routes
☐ Temporary Flight Restriction Areas

Uncontrolled Airspace

When you fold in considerations such as VFR vs. IFR in VMC and VFR vs. IFR in IMC, plus the wide variety of aircraft navigating the atmosphere, you can see you must (as we said at the beginning of this chapter) keep your wits about you.

CONGESTED AIRSPACE QUIZ

To help you make sure you are absorbing the material in this book (it's life or death data, so you *need* to), each chapter will end with a brief quiz. Answers are on page 181.

1. In which of the following types of airspace is a transponder mandatory?
   a. All TCAs.
   b. Group I TCAs only.
   c. Group II TCAs only.
   d. None of the above.

2. In which of the following types of airspace is an altitude encoder (transponder with Mode C) mandatory?
   a. All TCAs.
   b. Group I TCAs only.
   c. Group II TCAs only.
   d. None of the above.

3. What is the speed limit (IAS) for flight of a piston-engine aircraft within an Airport Traffic Area?

a. 150 knots or 172 mph.
b. 130 knots or 150 mph.
c. 156 knots or 180 mph.
d. 200 knots or 230 mph.

4. Special VFR weather minima for a light single- or twin-engine aircraft are:
a. Clear of clouds and one nautical mile visibility.
b. 500 feet below clouds and one statute mile ground visibility.
c. Published alternate minima.
d. Clear of clouds and one statute mile flight visibility.

5. Pilot and aircraft need *not* be legal for IFR flight when operating:
a. SVFR at night.
b. SVFR in Group I TCAs.
c. IMC in uncontrolled airspace.
d. Above FL 180.

6. Military aircraft are restricted to a 250 knot speed limit below 10,000 feet:
a. At all times.
b. In MOAs only.
c. In MTRs only.
d. None of the above.

7. Responsibility for collision avoidance between civilian and military aircraft within a Military Training *VR* is:
a. That of the pilot of the military aircraft only.
b. Shared by the military pilots and anyone else operating within the MTR.
c. That of the pilot of the civilian aircraft only.
d. That of ATC.

8. Which of the following are generally *not* marked on aeronautical charts?
A. Control Zones
B. Controlled Firing Areas

C. Airport Traffic Areas
D. MTRs
E. MOAs
F. Alert Areas
a. A, D, E & F.
b. B & C.
c. B, C & D.
d. C & F.

9. MTRs are depicted on:
a. Sectional charts only.
b. Low-altitude enroute charts only.
c. The "Green Demon" only.
d. All of the above.

10. For which airspace is it a good practice to obtain a clearance before penetration?
a. Prohibited Areas.
b. Restricted Areas and Warning Areas.
c. Alert Areas and Prohibited Areas.
d. Alert, Restricted and Controlled Firing Areas.

Chapter 2
Safety Rules

Safety rules are specified in the Federal Aviation Regulations (FARs), Part 91. Here are the most important "Rules of the Air." These are not complete. For the complete text, consult a current set of FARs. The order in which they appear here is based on logic, not on the illogical FAR numbering system.

91.7 CARELESS OR RECKLESS OPERATION

No person may operate an aircraft in a careless or reckless manner so as to endanger the life or property of another.

91.65 OPERATING NEAR OTHER AIRCRAFT

(a) No person may operate an aircraft so close to another aircraft as to create a collision hazard.

(b) No person may operate an aircraft in formation flight except by arrangement with the pilot in command of each aircraft in the formation.

91.67 RIGHT OF WAY (except water operations)

(a) General. When weather conditions permit, regardless of whether an operation is conducted under IFR or VFR, vigilence shall be maintained by each person operat-

Only by pre-arrangement may you do this. F33 and A36 Bonanzas. (courtesy Beech)

ing an aircraft so as to see and avoid other aircraft. When a rule of this section gives another aircraft the right of way, he shall give way to that aircraft, and may not pass over, under, or ahead of it, unless well clear.

(b) Distress. An aircraft in distress has the right of way over all other traffic.

(c) Converging. When aircraft of the same category are converging at approximately the same altitude (except headon, or nearly so) the aircraft to the other's right has the right of way. If the aircraft are of different categories:

- ☐ A balloon has the right of way over any category of aircraft.
- ☐ A glider has the right of way over an airship, airplane or rotorcraft.
- ☐ An airship has the right of way over an airplane or rotorcraft.
- ☐ An aircraft towing or refueling other aircraft has the right of way over all other engine-driven aircraft.

(d) Approaching head-on. When aircraft are approaching each other head-on, or nearly so, each pilot shall alter course to the right.

(e) Overtaking. Each aircraft that is being overtaken has the right of way, and each pilot of an overtaking aircraft shall alter course to the right to pass well clear.

(f) Landing. Aircraft while on final approach to land, or while landing, have the right of way over other aircraft in flight or operating on the surface. When two or more aircraft are approaching an airport for the purpose of landing, the aircraft at the lower altitude has the right of way, but shall not take advantage of this rule to cut in front of another aircraft which is on final approach to land, or to overtake that aircraft.

91.70 AIRCRAFT SPEED

(a) No person may operate an aircraft below 10,000

If you see this, get out of the way—unless you're in a glider or balloon! (courtesy Goodyear)

feet (except when otherwise authorized) at speed of more than 250 knots (288 mph) IAS.

(b) No person may operate an aircraft within an airport traffic area (except when otherwise authorized) at a speed of more than:

- ☐ Piston-engine aircraft—156 knots (180 mph) IAS.
- ☐ Turbine-powered aircraft—200 knots (230 mph) IAS.

(c) No person may operate an aircraft in the airspace underlying a TCA, or in a VFR corridor through a TCA at a speed of more than 200 knots (230 mph) IAS.

However, if the minimum safe airspeed for any particular operation is greater than the maximum speed prescribed in ths section, the aircraft may be operated at that minimum speed.

91.71 ACROBATIC FLIGHT

No person may operate an aircraft in acrobatic flight:

(a) Over any congested area of a city, town or settlement.

(b) Over an open-air assembly of persons.

(c) Below an altitude of 1,500 feet above the surface.

(d) When flight visibility is less than 3 miles.

91.73 AIRCRAFT LIGHTS

No person may, during the period from sunset to sunrise:

(a) Operate an aircraft unless it has lighted position lights.

(b) Operate an aircraft, required by FAR 91.33 to be equipped with anti-collision lights, unless it has approved and lighted aviation red or aviation white anti-collision lights. The pilot may keep them unlit if it is determined that it is safer to keep them off (e.g., in cloud).

91.79 MINIMUM SAFE ALTITUDES; GENERAL

Except when necessary for takeoff or landing, no person may operate an aircraft below the following altitudes:

(a) Anywhere. An altitude allowing, if a power unit fails, an emergency landing without undue hazard to persons or property on the surface.

(b) Over congested areas. Over any congested area of a city, town, or settlement, or over an open-air assembly of persons, an altitude of 1,000 feet above the highest obstacle within a horizontal radius of 2,000 feet of the aircraft.

(c) Over other than congested areas. An altitude of 500 feet above the surface, except over open water or sparsely populated areas. In that case, the aircraft may not be operated closer than 500 feet to any person, vessel, vehicle, or structure.

(d) Helicopters. Helicopters may be operated at less than the minimums prescribed in paragraph (b) or (c) of this section if the operation is conducted without hazard to persons or property on the surface. In addition, each person operating a helicopter shall comply with routes or altitudes specifically prescribed for helicopters by the Administrator.

91.109 VFR CRUISING ALTITUDE OR FLIGHT LEVEL

Except while holding in a holding pattern of 2 minutes or less, or while turning, each person operating an aircraft under VFR in level cruising flight more than 3,000 feet above the surface shall maintain the appropriate altitude or flight level prescribed below, unless otherwise authorized by ATC:

(a) When operating below 18,000 feet MSL and—

(1) On a magnetic course of zero degrees through 179 degrees, any odd thousand foot MSL altitude plus 500 feet (such as 3,500, or 5,500 or 7,500); or

(2) On a magnetic course of 180 degrees through 359 degrees, any even thousand foot MSL altitude plus 500 feet (such as 4,500, 6,500, 8,500).

(b) When operating above 18,000 feet MSL to flight level 290 (inclusive), and—

(1) On a magnetic course of zero degrees through 179 degrees, any odd flight level plus 500 feet (such as 195, 215, or 235); or

(2) On a magnetic course of 180 degrees through 359 degrees, any even flight level plus 500 feet (such as 185, 205, or 225).

(c) When operating above flight level 290 and—

(1) On a magnetic course of zero degrees through 179 degrees, any flight level, at 4,000-foot intervals, beginning at and including flight level 300 (such as flight level 300, 340, or 380); or

(2) On a magnetic course of 180 degrees through 359 degrees, any flight level at 4,000 foot intervals, beginning at and including flight level 320 (such as flight level 320, 360, or 400).

91.119 MINIMUM ALTITUDES FOR IFR OPERATIONS

(a) Except when necessary for takeoff or landing, or unless paragraph (b) of this section, no person may operate an airplane under IFR below—

(2) (i) In the case of operations over an area designated as a *mountainous area* in Part 95, an altitude of 2,000 feet above the highest obstacle within a horizontal distance of 5 statute miles from the course to be flown, or

(ii) In any other case, an altitude of 1,000 feet above the highest obstacle within a horizontal distance of 5 statute miles from the course to be flown.

However, if both a MEA (minimum enroute altitude) and a MOCA (minimum obstruction clearance altitude) are prescribed for a route, a person may operate an aircraft below the MEA down to, but not below, the MOCA, when within 25 statute miles of the VOR concerned (based on the pilot's reasonable estimate of that distance).

(b) Climb. Climb to a higher minimum IFR altitude shall begin immediately after passing the point beyond which that minimum altitude applies, except that, when ground obstacles intervene, the point beyond which the higher minimum altitude applies shall be crossed at or above the applicable MCA (minimum crossing altitude).

91.121 IFR CRUISING ALTITUDE OR FLIGHT LEVEL

(a) In controlled airspace. Each person operating an aircraft under IFR in level cruising flight in controlled airspace shall maintain the altitude or flight level assigned that aircraft by ATC. However, if the ATC clearance assigns "VFR conditions on-top," he shall maintain an altitude or flight level as prescribed by FAR 91.109.

(b) In controlled airspace. Except while holding in a holding pattern of two minutes or less, or while turning, each person operating an aircraft under IFR in level cruising flight, in uncontrolled airspace, shall maintain an appropriate altitude as follows:

(1) When operating below 18,000 feet MSL and—

(i) On a magnetic course of zero degrees through 179 degrees, any odd thousand foot MSL altitude (such as 3,000, 5,000, or 7,000); or

(ii) On a magnetic course of 180 degrees through 359 degrees, any even thousand foot MSL altitude (such as 2,000, 4,000, or 6,000).

(2) When operating at or above 18,000 feet MSL but below flight level 290, and—

(i) On a magnetic course of zero degrees through 179 degrees, any odd flight level (such as 190, 210, or 230); or

(ii) On a magnetic course 180 degrees through 359 degrees, any even flight level (such as 180, 200, or 220).

(3) When operating at flight level 290 and above, and—

(i) On a magnetic course of zero degrees through 179 degrees, any flight level, at 4,000-foot intervals, beginning at and including flight level 290 (such as flight level 290, 330, or 370); or

(ii) On a magnetic course of 180 degrees through 359 degrees, any flight level, at 4,000-foot intervals, beginning at and including flight level 310 (such as flight level 310, 350, or 390).

91.75 COMPLIANCE WITH ATC CLEARANCE AND INSTRUCTIONS

(a) When an ATC clearance has been obtained, no pilot in command may deviate from that clearance, except in an emergency, unless he obtains an amended clearance. However, except in positive controlled airspace, this paragraph does not prohibit him from cancelling an IFR flight plan if he is operating in VFR weather conditions. If a pilot is uncertain of the meaning of an ATC clearance, he shall immediately request clarification from ATC.

(b) Except in an emergency, no person may, in an area in which air traffic control is exercised, operate an aircraft contrary to an ATC instruction.

(c) Each pilot in command who deviates, in an emergency, from an ATC clearance or instruction shall notify ATC of that deviation as soon as possible.

(d) Each pilot in command who (though not deviating from a rule of this subpart) is given priority by ATC in an emergency, shall, if requested by ATC, submit a detailed report of that emergency within 48 hours to the chief of that ATC facility.

91.85 OPERATING ON OR IN THE VICINITY OF AN AIRPORT; GENERAL RULES

(b) Unless otherwise authorized or required by ATC, no person may operate an aircraft within an aircraft traffic area except for the purpose of landing at, or taking off from, an airport within that area. ATC authorizations may be given as individual approval of specific operations or may be contained in written agreements between airport users and the tower concerned.

91.87 OPERATION AT AIRPORTS WITH OPERATING CONTROL TOWERS

(a) General. Unless otherwise authorized or required by ATC, each person operating an aircraft to, from, or on an airport with an operating control tower shall comply with the applicable provisions of this section.

(b) Communications with control towers operated by the United States. No person may, within an airport traffic area, operate an aircraft to, from, or on an airport having a control tower operated by the United States unless two-way radio communications are maintained between that aircraft and the control tower. However, if the aircraft radio fails in flight, he may operate that aircraft and land if weather conditions are at or above basic VFR weather minimums, he maintains visual contact with the tower, and he receives a clearance to land. If the aircraft radio fails while in flight under IFR, he must comply with FAR 91.127 (IFR Radio Failure Rules).

(c) Communications with other control towers. No person may, within an airport traffic area, operate an aircraft to, from, or on an airport having a control tower that is operated by any person other than the United States unless—

(1) If that aircraft's radio equipment so allows, two-way radio communications are maintained between the aircraft and the tower; or

(2) If that aircraft's radio equipment allows only reception from the tower, the pilot has the tower's frequency monitored.

(d) Minimum altitudes. When operating to an airport with an operating control tower, each pilot of—

(1) A turbine-powered airplane or a large airplane shall, unless otherwise required by the applicable distance-from-cloud criteria, enter the airport traffic area at an altitude of at least 1,500 feet above the surface of the airport and maintain at least 1,500 feet within the airport traffic area, including the traffic pattern, until further descent is required for a safe landing;

(2) A turbine-powered airplane or a large airplane approaching to land on a runway being served by an ILS, shall, if the airplane is ILS equipped, be flown at an altitude at or above the glideslope between the outer marker (or the point of interception with the glideslope, if compliance with the applicable distance-from-clouds criteria requires interception closer in) and the middle marker; and,

(3) An airplane approaching to land on a runway served by a visual approach slope indicator shall maintain an altitude at or above the glideslope until a lower altitude is necessary for a safe landing.

However, subparagraphs (2) and (3) of this paragraph do not prohibit normal bracketing maneuvers above or below the glideslope that are conducted for the purpose of remaining on the glideslope.

(e) Approaches. When approaching to land at an airport with an operating control tower, each pilot of—

(1) An airplane shall, circle the airport to the left, and

(2) A helicopter, shall avoid the flow of fixed-wing aircraft.

(f) Departures. No person may operate an aircraft taking off from an airport with an operating control tower except in compliance with the following:

(1) Each pilot shall comply with any departure procedures established for that airport by the FAA.

(2) Unless otherwise required by the departure procedure or the applicable distance from clouds criteria, each pilot of a turbine-powered airplane and each pilot of a large airplane shall climb to an altitude of 1,500 feet above the surface as rapidly as practicable.

(g) Noise abatement runway system. When landing or taking off from an airport with an operating control tower, and for which a formal runway-use program has been established by the FAA, each pilot of a turbine-powered airplane and each pilot of a large airplane, assigned a noise abatement runway by ATC, shall use that runway. However, consistent with the final authority of the pilot in command concerning the safe operation of the aircraft as prescribed in FAR 91.3(a), ATC may assign a different runway if requested by the pilot in the interest of safety.

(h) Clearance required. No person may, at any airport with an operating control tower, operate an aircraft on a runway or taxiway, or takeoff or land an aircraft, unless an appropriate clearance is received from ATC. A clearance to "taxi to" the takeoff runway assigned to the aircraft is not a clearance to cross that assigned takeoff runway, or to taxi on that runway at any point, but is a clearance to cross other runways that intersect the taxi route to that assigned takeoff runway. A clearance to "taxi to" any point other than an assigned takeoff runway is a clearance to cross all runways that intersect the taxi route to that point.

91.115 ATC CLEARANCE AND FLIGHT PLAN REQUIRED FOR IFR

No person may operate an aircraft in controlled

airspace under IFR unless—

(a) He has filed an IFR flight plan; and

(b) He has received an appropriate ATC clearance.

91.123 COURSE TO BE FLOWN WHEN IFR

Unless otherwise authorized by ATC, no person may operate an aircraft within controlled airspace, under IFR, except as follows;

(a) On a Federal airway, along the centerline of that airway.

(b) On any other route, along the direct course between the navigational aids or fixes defining that route. However, this section does not prohibit maneuvering the aircraft to pass well clear of other air traffic or the maneuvering of the aircraft, in VFR conditions, to clear the intended flight path before and during climb or descent.

91.21 FLIGHT INSTRUCTION; SIMULATED INSTRUMENT FLIGHT AND CERTAIN FLIGHT TESTS

(a) No person may operate a civil aircraft (except a manned free balloon) that is being used for flight instruction unless that aircraft has fully functioning, dual controls.

How to avoid midair—do it on the ground! (courtesy ATC)

However, instrument flight instruction may be given in a single-engine airplane equipped with a single, functioning throwover control wheel, in place of fixed, dual controls of the elevator and ailerons, when:

(1) The instructor has determined that the flight can be conducted safely; and

(2) The person manipulating the controls has at least a private pilot certificate with appropriate category and class ratings.

(b) No person may operate a civil aircraft in simulated instrument flight unless—

(1) An appropriately rated pilot occupies the other control seat as a safety pilot;

(2) The safety pilot has adequate vision forward and to each side of the aircraft, or a competent observer in the aircraft adequately supplements the vision of the safety pilot; and

(3) Except in the case of lighter-than-air aircraft, that aircraft is equipped with fully functioning dual controls. However, simulated instrument flight may be conducted in a single-engine airplane, equipped with a single, functioning, throwover control wheel, in place of fixed, dual controls of the elevator and ailerons, when—

(i) The safety pilot has determined that the flight can be conducted safely; and

(ii) The person manipulating the control has at least a private pilot certificate with appropriate category and class ratings.

(c) No person may operate a civil aircraft that is being used for a flight test for an airline transport pilot certificate or a class or type rating on that certificate, or for an FAR Part 121 proficiency flight test, unless the pilot seated at the controls, other than the pilot being checked, is fully qualified to act as pilot in command of the aircraft.

SAFETY RULES SUMMARY

The foregoing has been a rather dry compendium of the most important FARs that apply to people who fly in congested airspace. In this chapter, we have reviewed some of the rules for:

- ☐ Operating aircraft
- ☐ Right of way
- ☐ Speed
- ☐ Acrobatic flight
- ☐ Aircraft lights
- ☐ Minimum safe altitudes
- ☐ Flight altitude rules
- ☐ VFR cruising altitudes
- ☐ Minimum altitudes for IFR operations
- ☐ IFR cruising altitudes
- ☐ Compliance with ATC clearances and instructions
- ☐ Operating at airports
- ☐ ATC clearance and flight plan required for IFR
- ☐ Course to be flown when IFR
- ☐ Flight instruction—simulated instrument flight

As you can see, there are many complex rules governing aircraft operations that are particularly applicable in congested airspace. It is a good idea to review these from time to time, since they are what the other guys up there are supposed to be guided by. To refresh your memory, take the accompanying quiz every six months or so.

SAFETY RULES QUIZ

Answers are on page 181.

1. You are VFR, approaching a VOR, tracking the 182° radial on a heading of 350°. You would be legal at an altitude of:
   a. 6,000 feet.
   b. 6,500 feet.
   c. 7,000 feet.
   d. 7,500 feet.

2. You see an aircraft traveling toward the same VOR, at your altitude. He is tracking in on the 270° radial. You appear to be on a collision course. Who has the right of way?
   a. You do.
   b. He does.
   c. Both shall alter course to the right.
   d. Both shall alter course to the left.

3. You are on close final approach to an uncontrolled airport. Without warning, an aircraft pulls out onto the runway you planned to land on and takes off right in front of you. You elect to go around. How should you overtake the other aircraft?

   a. Stay to his right.
   b. Maintain altitude and overfly him.
   c. Stay to his left.
   d. Do a 360° turn to the right.

4. You are on your way to the major airport in the area, and are receiving traffic advisories from its radar approach controller. As you are being vectored to the destination airport, you notice that you are overflying another controlled airport at 2,000 feet AGL. You should:

   a. Contact the other airport tower and obtain a clearance through their control zone.
   b. Talk to the other tower on your second radio, and let him know you are overflying him, while continuing to communicate with the radar controller.
   c. Advise the radar controller you are overflying the airport, and request a clearance to go through the control zone.
   d. Continue on your vector, maintain communication with the radar controller, and watch out for other aircraft.

5. The ground controller clears you to "Taxi to runway 27R." You may, without further clearance:

   a. Taxi across any other runway or taxiway to get to 27R, but you may not enter 27R.
   b. Taxi across runway 27R, if you need to, to get to the threshold.
   c. Not cross any other runway.
   d. Not cross any other taxiway.

6. The speed limits below 10,000 feet and beneath a TCA are:

   a. The same.
   b. 250 knots and 200 knots respectively.

c. 200 knots and 250 knots respectively.
d. 250 knots and 156 knots respectively.

7. To fly IFR in controlled airspace in the U.S., you need:
a. An instrument rating, a flight plan and a clearance.
b. An instrument rating and a clearance.
c. "a" plus an aircraft equipped with a transponder and DME.
d. "b" plus an aircraft equipped with a transponder and DME.

8. Which aircraft has the right of way?
a. Balloon over gyrocopter.
b. Gyrocopter over airship.
c. Airship over balloon.
d. Airship over aircraft towing a glider.

9. A TCA which you wish to overfly from the southeast has an upper limit of 7,000 feet. What is the minimum legal VFR altitude at which you may do this?
a. 7,000 feet
b. 7,500 feet
c. 8,000 feet
d. 8,500 feet

10. A pilot may *not* "cancel IFR" when:
a. In positive control airspace.
b. In a TCA.
c. Flying at night in VMC.
d. Waiting for a clearance after a flight plan has been filed.

Chapter 3
Preparing Yourself

You've had an opportunity to review the types of congested airspace and some of the FARs that affect operations when you fly in it. Now let's take a look at how you should be prepared, as a pilot, to handle it. Here are the considerations:

- ☐ Your qualifications
- ☐ Your level of training
- ☐ Your experience
- ☐ Your airplane's equipment
- ☐ Charts
- ☐ Where you are going

YOUR QUALIFICATIONS

It's a good idea to go for the highest qualifications you can. Not only are you likely to get a lower insurance rate, but the training you have to go through to become qualified can only help you deal with congested airspace.

The Instrument Rating

Without question, the best qualification you can get to fly in congested airspace is an instrument rating. This is because training for an instrument rating puts you into that part of the airspace system where the professionals fly every day (also known as *congested airspace*). Professionals *have* to be good. Constant exposure to this system will develop your own abilities and level of confidence.

To obtain an Instrument Rating, you must have a Private or Commercial Pilot Certificate, meet certain experience requirements, pass a written exam and a flight test. The experience requirements are:

- ☐ Total time 200 hours.
- ☐ Pilot in command (PIC) time 100 hours.
- ☐ PIC cross-country time 50 hours.
- ☐ Total simulated or actual instrument time 40 hours.
- ☐ Instrument instruction time (included in above) 20 hours.
- ☐ Instrument instruction time in the air (included) 5 hours.

Written Tests

You must pass a multiple-choice written exam for the instrument rating. The FAA publishes all the questions—thousands of them—in Advisory Circulars, which you can buy at most fixed base operators (FBOs) or government book stores. When you take the exam, you get a booklet with all the questions and another booklet telling you which questions you have to answer—maybe a hundred or so of the thousand or so in the booklet. Before you take any written exam, you must supply proof that you have completed the ground instruction or home study course required, and you must also show indentification and proof of age.

An FAA exam is good for two years—in other words, you must complete everything else pertaining to the certificate (the flying time and flight test)—within two years of taking the written.

Tests are usually taken at the local FAA General Aviation District Office (GADO), or at the facilities of your flight school. If you are taking the test at a GADO, be sure you allow yourself enough time to complete the test before they go home. Quitting time is usually 4:30 p.m., so if you have a four-hour exam, you must get there before 12:30 p.m. or they won't let you take it.

YOUR LEVEL OF TRAINING

The act of obtaining an instrument rating gives you an excellent level of training. But then what? You have to keep proficient. This means constant practice. You'll find the skills you so carefully honed for your instrument flight test will leave you very quickly (especially when your overall experience is low) if you don't keep at it.

YOUR EXPERIENCE

If you don't have an instrument rating, you have even more reason to keep practiced up. I'm not talking about takeoffs and landings; I'm not talking about radio navigation procedures and communications. Take every opportunity to fly in controlled-airspace situations—especially if they're not too congested—so you can keep current on procedures and practices in your area. If you are *really* in the dark about it, find another pilot who is experienced and ask if you can go along on a trip, or ask him or her to accompany you on one of your trips.

VISITING ATC FACILITIES

A very valuable experience for you is to visit the various FAA facilities. These include Flight Service Stations (FSS), Control towers, Radar approach controls (RAPCON), and ATC Centers. In all cases, pick an unbusy time, call the nearest facility, identify yourself as a pilot, and ask if you can visit for a while.

Flight Service Stations

At the FSS, ask about your local airspace—who controls what, where the radar facilities and special airspace areas (e.g., restricted, prohibited, military operations areas) are, and so on. Give the specialist a typical route you might take, and ask him or her to take you through it from these points of view. Ask to see how the computer-retrieval system works for getting data on military training routes, Notams, and so on. Find out how a flight plan is filed, and what they do with it. Ask about the different radio frequencies they use. For example, many FSS can talk to you on the frequencies of nearby VORs, in addition to broadcasting weather information over these stations.

Control Towers

When you go to the tower, first pause to catch your breath! It's often a steep climb up a spiral staircase to get to the cab. Bear in mind that the controllers are there to control traffic, not show you around, so be inconspicuous and pick your moments to ask questions carefully.

At the tower, you'll usually find at least a ground controller who handles aircraft and traffic moving on the surface of the airport, and a local controller who talks to aircraft in the air, landing, and those about to take off. You may also find a supervisor and perhaps a coordinator, who deals with things like receiving flight plans, talking to other FAA facilities, answering the telephone, and so on.

Ask to see the radar, if they have it, and see how they coordinate with other control facilities. Watch several aircraft coming in, preferably on instrument approaches. Find

Wichita tower. (courtesy Beech)

out about local landmarks used to report airplane positions. Ask what their biggest problems are.

Radar Approach Controls

The RAPCON is a very quiet, dark room, with people huddled over glimmering screens. Standing behind these are supervisors and coordinators, who interact between various pieces of airspace, and with the area's control towers and ATC center. It is a place well worth a visit. Explain the types of trips you make, and find out how you would be handled. Ask to see a non-transponder radar target, one that is "identing," and how the data blocks work on the radar screen. Watch the approach controller for a while, and then the departure controller. Once again, be unobtrusive, and be ready to get out of the way or stop talking if something happens to take your host's attention away from you.

ATC Centers

Most centers are away from airports—sometimes way out in the boondocks. These are the ATC centers, as of 1982:

- ☐ Albuquerque
- ☐ Atlanta
- ☐ Boston
- ☐ Chicago
- ☐ Cleveland
- ☐ Denver
- ☐ Fort Worth
- ☐ Houston
- ☐ Indianapolis
- ☐ Jacksonville
- ☐ Kansas City
- ☐ Los Angeles
- ☐ Memphis
- ☐ Miami
- ☐ Minneapolis
- ☐ New York
- ☐ Oakland
- ☐ Salt Lake City

- ☐ Seattle
- ☐ Washington

When you visit a Center you'll probably be given a tour of the various positions that cover the sectors for which the Center is responsible. Once again, you'll have plenty of opportunities to ask questions about how your type of flying fits in with their operation.

YOUR AIRPLANE'S EQUIPMENT

The better the equipment the airplane you fly has, the easier it will be for you to fly in congested airspace. Today's avionics have many labor-saving features that are well worth having. For example, many navigational radios (NAVs) and communications sets (COMMs) have a frequency storage system that lets you dial in your next frequency in advance of need, then summon it for use at the touch of a button. Some new NAVs feature digital VOR readouts, or you can buy a digital VOR indicator that will connect to your NAV unit. Digital VOR is very useful, since it always reads your current bearing to or from the station, eliminating the need to twist knobs when you want fast answers. Distance Measuring Equipment (DME) is also a useful aid. This gives your distance in nautical miles from most VORs.

The author's Comanche has a highly customized panel.

King KNS80 RNAV. (courtesy King).

I strongly recommend an area navigation system (RNAV). The most common systems use VOR and DME signals to compute position information. I have used a King KNS 80 in my Comanche for several years, and would not have an airplane without it. RNAV enables you to relocate any VORTAC to any position within about 200 miles of the station. Thus you can fly almost anywhere as easily as if you were going direct to a VOR. Since a lot of the flying you do in congested areas requires changes in routings, an RNAV that can be easily reprogrammed is very handy. One of the best features of RNAV is that it tells you how *far* you are from your programmed destination (called a waypoint). The better units also give your ground speed to the waypoint, and tell you how long it will take to get there.

Of course, a transponder is essential if you are to fly in congested airspace. Today's ATC radars are designed for use with transponders. To fly in congested airspace without a transponder is asking for trouble. An altitude encoder is also highly recommended. You *must* have one to fly in Group I TCAs. These are at Atlanta, Boston, Chicago, Dallas/Fort Worth, Los Angeles, Miami, New York San Francisco and Washington, D.C.

I strongly recommend a headset and boom microphone, with a push-to-talk button on the control wheel. You can buy portable units that will work on rented aircraft. With this setup you eliminate the need for that third hand you so often seem to need when things get busy.

Here is the ideal minimum avionics fit for flying in high density traffic areas:

This you need in busy areas. (courtesy David Clark)

- ☐ 360 channel COMM (prefer with frequency preselect)
- ☐ VOR (prefer with digital VOR readout)
- ☐ Transponder
- ☐ Encoder
- ☐ Headset and boom mike

In addition, if finances permit, I would add:

- ☐ Second COMM
- ☐ Second VOR
- ☐ DME and RNAV

CHARTS

Having a good set of charts will make life a lot easier as you fly in congested areas, and reviewing them in advance of a flight is a good idea. You can get aviation charts for VFR pilotage by visual or radio navigation, for IFR enroute navigation, and for IFR approaches and departures at airports. They are available from government and commercial sources. All U.S. Government charts are published at regular intervals by the National Ocean Survey (NOS). All their charts can be bought by subscription.

Visual Charts

NOS visual charts come in several formats. The two most commonly used are the Sectional and VFR Terminal Area Charts. Sectionals have a scale of 1:500,000—about eight miles to the inch. VFR Terminal Area Charts have a scale of 1:250,000—about four miles to the inch. These can be particularly useful, since they cover most of the congested areas:

- ☐ Atlanta
- ☐ Boston
- ☐ Chicago
- ☐ Cleveland
- ☐ Dallas/Fort Worth
- ☐ Denver
- ☐ Detroit
- ☐ Houston
- ☐ Kansas City
- ☐ Las Vegas
- ☐ Los Angeles

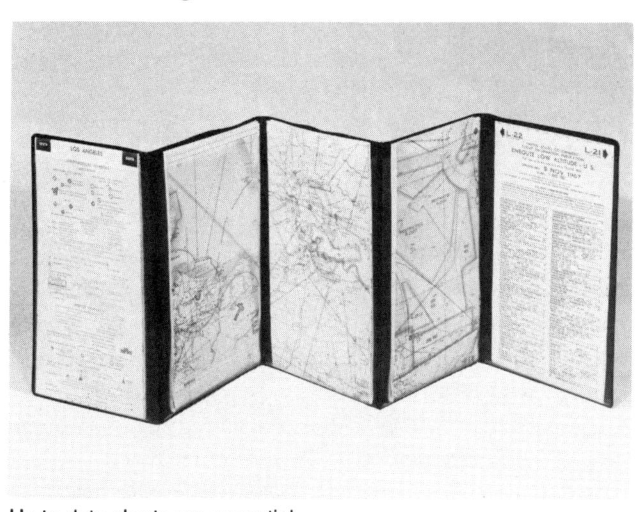

Up-to-date charts are essential.

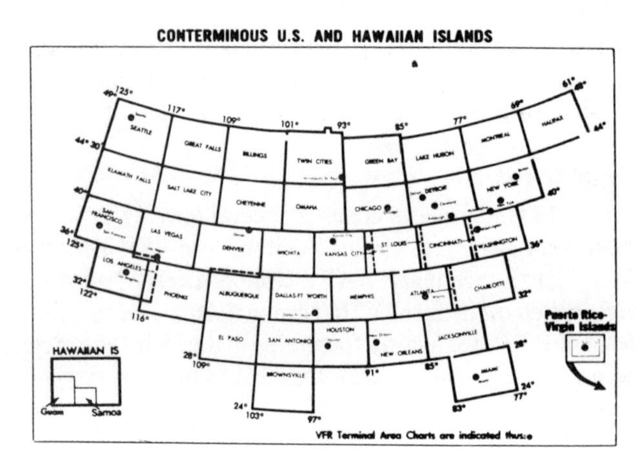

Charts—Sectional and VFR Terminal Areas.

- ☐ Miami
- ☐ Minneapolis/St. Paul
- ☐ New Orleans
- ☐ New York
- ☐ Philadelphia
- ☐ Pittsburgh
- ☐ San Francisco
- ☐ Seattle
- ☐ St. Louis
- ☐ Washington

Sectional and Terminal Area Charts are available over the counter at most airport dealers, or by subscription.

Another series of VFR visual/radio navigational charts, called Flightcharts, is available. These come in subscriptions covering six broad areas of the United States, and use a variety of scales, like radio-facility charts (this is an important difference from Sectionals, which have a constant scale). Flightcharts are reissued every six months. The main benefits of Flightcharts are that the VOR navigational information is easier to read than on a Sectional, and they contain more airport information.

Sky Prints is an atlas of VFR radio navigation charts, with no topographical information. It shows direct VOR-to-VOR courses as well as airways. It's issued annually, with a monthly information-update sticker.

IFR Charts

NOS publishes a complete set of IFR enroute and approach charts. The approach charts are issued in bound books and reissued every 56 days. These may be ordered on a subscription basis. You can also buy charts as a one-shot deal, which is useful for special trips.

Jeppesen publishes its own set of charts, and provides worldwide coverage on a customized subscription basis.

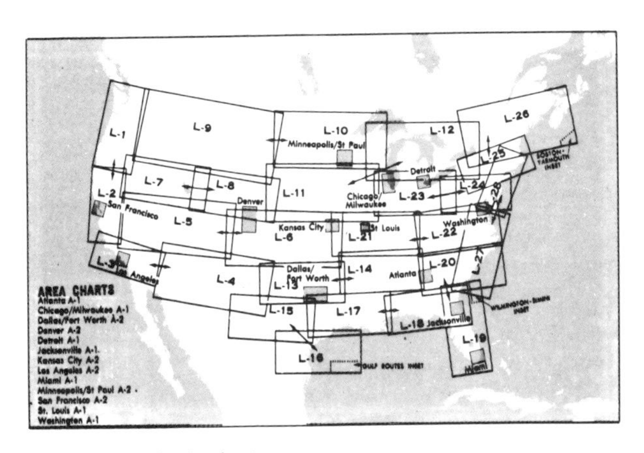

Enroute Low Altitude charts.

Jepp charts are issued in loose-leaf binders and are revised every week. Jeppesen also issues a VFR/IFR RNAV (area navigation) chart series. If you are a Jeppesen subscriber, you can order a "trip-kit" for an unusual trip, with no renewal service.

Charts may be ordered through the various commercial aviation catalogs, direct from the producers, or through the Aircraft Owners and Pilots Association.

WHERE YOU ARE GOING?

This brings us to my exclusive CACL system. CACL means *Congested Airspace Checklist.* By running through this before a trip, you should eliminate most of the potential problems before they can occur. Here it is:

CONGESTED AIRSPACE CHECKLIST

Destination ________________ Date ______________
Route ______________________ Aircraft ____________

Factors to consider for this route:

TCAs ______________________________________
Control Zones ______________________________
Prohibited Areas ____________________________
Restricted Areas ____________________________
Warning Areas ______________________________
Alert Areas ________________________________
Temporary Restricted Areas ___________________
Military Operations Areas ____________________
Military Training Routes ______________________
Sectional/Terminal charts needed ______________
Enroute charts needed _______________________
Airport charts needed _______________________

CACL will be of value to you in the earliest days of flying in congested areas. Later, when you become more accomplished, you need not bother with it, since you'll cover the points automatically. To use CACL, make a photocopy of the checklist and get out a VFR topographical chart. Look at your planned route and run down the CACL, filling out the answers to all the questions. When you enter data about Restricted Areas, note the controlling agency.

Call your FSS and get a weather briefing. Then ask for a rundown on NOTAMS, and ask if there are any Military Training Routes or other special airspace restrictions.

Get out all the charts you need—Sectionals, Terminals, Enroutes and Airports—and review them. Become familiar with their layout, so you can look up frequencies quickly. Your objective is to reduce head-down time when you're flying.

This chapter has discussed the steps you should take to prepare yourself for flight as pilot in command in congested areas. To make sure you've absorbed it, try the quiz on this page.

PREPARATION QUIZ

Answers are on page 181.

1. To obtain an instrument rating, you must have at least:

a. 200 hours pilot in command time.
b. 20 hours actual instrument time.
c. 20 simulated instrument time.
d. 200 hours total time.

2. RNAV enables you to:

a. Navigate "direct" anywhere in the U.S.
b. Artificially relocate VORTACs elsewhere within their range.
c. Fly legally within TCAs.
d. Dispense with a DME.

3. VFR Terminal Area Charts have a scale of:

a. 4 miles to the inch.
b. 1:250,000
c. 8 miles to the inch.
d. 1:500,000

4. An encoding altimeter is mandatory in:

a. Group I TCAs.
b. Group I and Group II TCAs.
c. All TCAs.
d. All controlled airspace.

5. You have passed an FAA written exam. You now have how much time to complete all other requirements for the certificate or rating?

a. Six months.
b. One year.
c. Two years.
d. Three years.

Chapter 4
Collision Avoidance

Let's face it: One of the main concerns you have in mixing with other airplanes is not bumping into them. Surprisingly, most collisions occur in excellent visibility in daylight.

According to a study by the National Transportation Safety Board (NTSB), there were 540 midair collisions involving general-aviation aircraft in the 20-year period from 1960 to 1979—an average of 27 a year, or about one every two weeks (Table 4-1). The accompanying table shows a breakdown of these accidents by year, and by type of aviation involvement—general aviation, air carrier or military (military vs. military collisions are not included.)

No less than 490 of these (90 percent) were between general aviation aircraft exclusively 18 (3 percent) were between general aviation and air-carrier aircraft. And 32 (6 percent) were between general aviation and military aircraft. Only four collisions in the 20-year period did not involve general aviation aircraft.

Given the high preponderence of general aviation flying, it's not surprising to see these statistics skew so heavily this way. And since that's the kind of flying you probably do, what are you going to do about it?

One thing you can do is practice collision avoidance techniques. This involves an awareness of risk, knowing

Table 4-1. Summary of Midair Collisions 1960-1979.

Year	Accidents Total	Fatal	Number Fatalities	GA vs GA	GA vs AC	GA vs MA	AC vs AC	AC vs MA
1960	26	10	152 a	19	4	2	1	0
1961	20	10	22	20	0	0	0	0
1962	19	9	27	14	0	5	0	0
1963	13	3	6	11	0	2	0	0
1964	15	7	12	13	0	2	0	0
1965	27	14	30	24	0	2	1	0
1966	27	11	33	25	1	1	0	0
1967	26	20	157	20	2	3	0	1
1968	37	23	69	33 b	3	1	0	0
1969	28	12	122	23 b	3	2	0	0
1970	37	21	55	32 b	0	5	0	0
1971	32	20	96	27	3	1	0	1
1972	25	13	41	24 b	1	0	0	0
1973	24	12	29	24	0	0	0	0
1974	34	19	48	32	0	2	0	0
1975	29	13	47	28	0	1	0	0
1976	31	24	64	30	0	1	0	0
1977	34	17	41	34	0	0	0	0
1978	35	23	189 c	33	1	1	0	0
1979	25	14	34	24	0	1	0	0
TOTAL	544	295	1,274	490	18	32	2	2

a Includes 6 persons on ground
b Includes 1 U.S. general aviation vs. foreign aircraft
c Includes 7 persons on ground
Source: National Transportation Safety Board 8-3-81.

*KEY: AC = Air Carrier Aircraft
GA = General Aviation Aircraft
MA = Military Aircraft

how to deal with this, knowing how to scan, where the best places are to look, and making the best use of ground radar for traffic advisories.

AREA OF GREATEST RISK

According to an article in *AOPA Pilot* magazine (November 1979):

- ☐ 45% of midair collisions occur outside the traffic pattern.
- ☐ Of midair collisions occurring during the takeoff or landing phase, 29% occurred at controlled airports.
- ☐ About 43% of midairs resulted in fatalities.
- ☐ Only 25% of midair collisions involved flight instruction or a solo student.

The greatest risk exposure is thus to be found at uncontrolled airports in the takeoff and landing phase.

AVOIDING COLLISIONS AT UNCONTROLLED AIRPORTS

Operating at an uncontrolled airport means you are strictly on your own as far as traffic avoidance is concerned. The problem is that there are a lot of undisciplined pilots who do anything they feel like at such fields. For example, they:

- ☐ Make no radio contact when arriving or departing.
- ☐ Take off from runways that are already in use by other aircraft.
- ☐ Cross runways that are in use.
- ☐ Backtrack down runways on which aircrafts are landing.
- ☐ Stay on the runway too long after landing.
- ☐ Cut other aircraft off in the pattern.
- ☐ Pull out onto the runway when another aircraft is on short final approach.

How do I know this? Because I have personally observed all of these infractions on numerous occasions at uncontrolled fields.

YOUR OWN SAFETY RULES

Here are some rules to observe when operating

around airports, particularly out of uncontrolled fields:

- ☐ Fly defensively! Assume that all other pilots will make mistakes and break rules.
- ☐ Don't make mistakes and don't break the rules!
- ☐ Follow the radio communications procedures suggested in Table 4-2.
- ☐ Constantly scan all around you for other aircraft.
- ☐ Look around before turning.
- ☐ Look around before descending.
- ☐ Be aware of the visual shortcomings of different aircraft, and take this into consideration when you know they are around.
- ☐ Maintain vigilance until the aircraft and propeller have both stopped.
- ☐ Use current charts.
- ☐ When you see your airplane's shadow on the ground, keep checking for a second shadow. It's better than radar!

Let's look into these rules in a little more depth.

Fly Defensively

Just as when you drive, assume the other guy is out to get you.

Obey the Rules Yourself

Do we really need to say this?

Follow Recommended Radio Communications Procedures

There are plenty of ways to get your message across at an uncontrolled airport. If there is no tower or FSS at the field, there is probably a Unicom station. This is the private aeronautical frequency—sort of an aviation party line—that is available at many airports. Unicom frequencies are standard at 122.7, 122.8, or 123.0. The specific frequency should be listed on the chart.

Bear in mind that the person at the other end may be a line boy or a receptionist behind the counter at the FBO—in other words, a person with little or no flying experience. Communications on Unicom include requests for wind and runway, advice that you are approaching or leaving, and specific position reports, which should be made when arriving on downwind and when turning final, and when departing as you taxi onto the runway for takeoff. In addition, you can use Unicom for brief messages, such as requests for taxi or hotel reservations, aircraft servicing, or "phone home."

Some airports have Unicom, but leave the volume turned down or have nobody monitoring it. In such a case, make blind transmissions of your location, as above. The biggest problem with Unicom is that the frequency gets very congested, especially on weekends. Just keep trying!

If the airport has no Unicom, tower, FSS or other frequency, broadcast your intentions and position reports on 122.9.

At some airports, the tower only operates limited hours. If the tower is temporarily closed, use the tower local control frequency for traffic advisories, unless another frequency is specified in an applicable NOTAM. If you broadcast your position or intentions in the blind, and a Unicom is in operation at the airport, you can get the wind direction and runway in use from Unicom, even though you use the tower frequency for self-announce procedures.

If there is an FSS open on the field, they probably operate an Airport Advisory Service (AAS), through which you make position reports using a dedicated FSS frequency. Your airport charts or the Airman's Information Manual should have information about this.

If the FSS is closed, broadcast your position or intentions in the blind on the published AAS frequency.

Refer to Table 4-2 for a cross-reference of suggested procedures.

Constantly Scan for Other Aircraft

It's all very well to advise you to look around all the time, but did you know that there's a technique for scanning? There is, and here it is.

First, bear in mind that you should make a frequent periodic scan of the entire area. Because of the performance characteristics of many aircraft, both in speed and

Table 4-2. Recommended Outbound and Inbound Reports—Non-controlled Airports.

FACILITY AT AIRPORT	FREQUENCY	BROADCAST POSITION	
		Outbound	Inbound
1. Unicom Operator (No Tower or FSS)	Communicate with Unicom operator on 122.7, 122.8, or 123.0 as appropriate. If unable to contact Unicom operator, use appropriate Unicom frequency to broadcast position or intentions in the blind.	Before taking runway for takeoff.	Entering downwind, and final.
2. Part-Time Tower* Closed, FSS Closed or no FSS	Broadcast position or intentions in the blind on tower frequency.	Before taking runway for takeoff.	Entering downwind, and final.
3. FSS Closed (No Tower)	Broadcast position or intentions in the blind on published Airport Advisory Service (AAS) frequency.	Before taking runway for takeoff.	Entering downwind, and final.

4. No Tower, FSS or Unicom Operator	Broadcast position or intentions in the blind on 122.9	Before taking runway for takeoff.	Entering downwind, and final.
5. Part-Time Tower* Closed, FSS Open	Communicate with FSS on tower frequency for Airport Advisory Service (AAS)	Before taxiing and taking runway for takeoff.	10 miles from airport, entering downwind and final.
5. FSS Open (No Tower)	Communicate with FSS on published Airport Advisory Service (AAS) frequency.	Before taxiing and taking runway for takeoff.	10 miles from airport, entering downwind and final.

*If tower temporarily closed use tower local control frequency for traffic advisory practices unless another frequency is specified in applicable NOTAM. If broadcast of position or intentions in the blind (self-announce) is used and a Unicom is in operation at the airport, it is suggested the wind direction and runway in use be obtained from UNICOM even though tower frequency is used for self-announce procedures.

climb or descent rates, an unwelcome stranger can be upon you in a few seconds. The high closure rates possible limit not only the time available for detection, but also how long you have to make a decision about the possible conflict and how long it actually takes for you to maneuver your airplane out of the way.

Clearly, the more time you spend looking for traffic, the greater the probability you will find a potential collision threat. But the time should be spent productively. You should use a system. Your eye always wants to focus on something. To be most effective, move your visual aim and refocus every so often. Don't just stare; work at it. Use short, regular eye movements to bring successive areas into your central visual field. Stay on each aiming point for at least one second, and change direction no more than ten degrees each time. Get comfortable with a pattern that you develop with constant practice.

Have you noticed how easy it is to detect movement out of the corner of your eye? Peripheral vision is very good at spotting motion, so when you shift your glance, pay attention to what your peripheral vision tells you. However, motion isn't really the key clue to a *dire* collision threat. That's because something that you are on a collision course with won't appear to be *moving* relative to you; it'll just be getting *larger*. So although apparent movement is often the first hint of a collision hazard, it's more likely the *hint* of a threat rather than a probability.

So what is most important is the ability to detect *no* relative movement between you and the other aircraft—just expansion in size. But bearing in mind that airplanes can be maneuvered in three dimensions, spotting an airplane that appears to be moving relative to you and that is nearby is still cause for concern and possible evasive action.

Night Flying

There's good news and bad news about night flying. First, it's often easier to spot traffic at night due to strobe lights, rotating beacons, etc. However, it's harder to see these against a background of city lights. And some aircraft just have steady red, green, and white position lights, making them very hard to see in any event.

The eye's ability to detect dim lights in the dark is better when you don't look directly at them. Your peripheral vision is better at detecting things that are faintly lit, so keep your eyes moving, and stop every few seconds to spot possible hazards.

Move Your Head

Bear in mind your airplane's visibility characteristics and fly it accordingly. Move your head occasionally to peer around obstructions such as door and window posts, struts, etc.

Look Around Before Turning

Remember when you were learning to fly? Remember how your instructor stressed the importance of taking a good look before making a turn? Have you done that lately? It's easy to get into bad habits, and not looking before turning is one of the worst. Take a good look before you turn (or carry out any maneuver that will result in a change in direction), paying particular attention to the airspace which you are going to be occupying shortly.

Look Around Before Descending

Same comments apply.

This Cessna Centurion has a radar pod to lift out of the way when the pilot is looking to the right. (courtesy Cessna)

Which do you think has better visibility—the high-wing Cessna 152 . . .? (courtesy Cessna)

Visual Shortcomings of Different Aircraft

With few exceptions, most light aircraft have a terrible visibility problem in at least one direction. Let's look at the main configurations and what their problems are.

High-wing aircraft are probably the most prevalent general aviation types around. When in a turn, they provide no visibility *into* the turn. When flying one, you can't see where you are going when you're turning. You also have a problem getting a good view above, and to the rear—even though many high-wing Cessnas have a rear and/or a roof window.

. . . or the low-wing Piper Tomahawk? (courtesy Piper)

Pilots of low-wing aircraft can't see down so well. However, when they turn, the wing drops out of the way, giving good visibility into the turn.

Conventional light twins have a big engine sitting out there on each wing, effectively blocking much of the view down and diagonally forward. They usually have good visibility directly ahead, however.

This Piper Seneca II, in the SFO TCA, demonstrates how those big engines can block vision in some directions. (courtesy Piper)

A Baby Lakes demonstrates the good and bad points of visibility. (courtesy Barney Oldfield)

Even though most biplanes have open cockpits that generally provide excellent visibility, the array of wings, struts, wires, etc., tends to cut down on visibility, giving the worst characteristics of both high and low-wing air-

A notorious example of bad visibility in a tailwheel aircraft—the Cessna 195. (courtesy Cessna)

craft. However, in many biplanes, the pilot sits behind the upper wing, which eliminates the bad turn visibility problem of a high-wing plane.

Most tailwheel-equipped aircraft have very poor visibility directly forward when landing and taxiing. In fact, when flared for touchdown, many taildraggers provide *zero* forward visibility.

The biggest visibility problem that occurs is when you get a mix of types flying in each other's blind spot. For example, probably the worst combination is a high-wing airplane on approach, with a low-wing single or light twin above and slightly to the rear, traveling in the same direction.

Maintain Vigilance Until the Aircraft and Propeller Have Both Stopped

It's easy to relax after you've landed and become a little careless as you taxi to the ramp. I know. I almost creamed a Gulfstream II one time at Washington National because I was hurrying off the runway at the controller's request. The G2 was on my left, taxiing to the same destination as was I. I was changing frequencies and not looking where I was going for a moment as I taxied quickly. We both braked hurriedly. That could have been expensive!

Use Current Charts

It's easy to get comfortable with an old pair of jeans or loafers. But don't do that with your charts! Charts are reissued on set cycles—usually every 56 days for IFR charts and every six months for visual charts. They contain frequency and name changes, as well as new hazards. As an example, the VOR near Wilmington, Delaware, airport had three different names and identification codes within one year: New Castle (EWT), Fatima (FYT) and to Dupont (DQO)! New Restricted and Prohibited Areas, variations in the limits of TCAs, new TV towers and airport closings are all examples of what may have happened that won't be on your old chart. Current charts are very important to your personal safety.

Use Your Airplane's Shadow

On a sunny day, use your airplane's shadow as a traffic advisory system. If it is alone, you're safe (unless there is an airplane between you and the sun). Yours is the shadow that always stays in the same place relative to your plane.

MAKE USE OF ATC RADAR

As a VFR aircraft, you can obtain radar assistance and navigation service (vectors) from radar-equipped FAA ATC facilities, provided you can communicate with the facility, are within radar coverage, and can be radar identified. Here are some notes from the FAA on the subject:

Radar Assistance

The FAA wants you to clearly understand that authorization to proceed in accordance with VFR radar-navigational assistance does not constitute authorization to violate FARs. In effect, assistance is provided on the basis that navigational-guidance information issued is advisory in nature and the job of flying the aircraft safely remains with you.

In many cases, the controller will be unable to determine if flight into instrument conditions will result from his instructions. To avoid possible hazards resulting from being vectored into IFR conditions, you should keep the controller advised of the weather conditions in which you are operating and along the course ahead.

Radar navigation assistance (vectors) may be initiated by the controller when one of the following conditions exists:

- ☐ The controller suggests the vector and the pilot concurs.
- ☐ A special program has been established and vectoring service has been advertised.
- ☐ In the controller's judgment the vector is necessary for air safety.

Radar navigation assistance (vectors) and other radar-derived information may be provided in response to your request. Many factors, such as limitations of radar, volume of traffic, communications-frequency congestion, and controller workload could prevent the controller from providing it. The controller has complete discretion for determining if he is able to provide the service in a particular case. His decision not to provide the service in a particular case is not subject to question.

Terminal Radar Programs for VFR Aircraft

Stage I Service (Radar Advisory Service for VFR Aircraft), in addition to the use of radar for the control of IFR aircraft, provides traffic information and limited vectoring to VFR aircraft on a workload-permitting basis. Vectoring service may be provided when requested by the pilot or with pilot concurrence when suggested by ATC. When arriving, you should contact approach control on the published frequency and give your position, altitude, radar-beacon code (if transponder equipped), destination, and request traffic information. Approach control will issue wind and runway information, except when you state "I have the numbers" or this information is contained in the ATIS broadcast and you indicate you have received the ATIS Information. Traffic information is provided on a workload-permitting basis. Approach control will specify the time or place at which you are to contact the tower on the local control frequency for further landing information. Upon being told to contact the tower, radar service is automatically terminated.

Stage II Service (Radar Advisory and Sequencing for VFR Aircraft) has been implemented at certain terminal locations. The purpose of the service is to adjust the flow of arriving VFR and IFR aircraft into the traffic pattern in a safe and orderly manner and to provide radar traffic information to departing VFR aircraft. Pilot participation is urged but it is not mandatory. When arriving VFR, you should initiate radio contact (frequencies listed in the AIM Airport Directory, NOS or Jeppesen charts) with approach control when approximately 25 miles from the airport at which Stage II services are being provided. On initial contact by VFR aircraft, approach control will assume that Stage II service is requested. Approach control will provide you with wind and runway (except when you state "Have the numbers" or that you have received the ATIS information), routings, etc., as necessary for proper sequencing with other participating VFR and IFR traffic enroute to the airport. Traffic information will be provided on a workload-permitting basis. If you don't want the service, you should state "Negative Stage II" or make a similar comment on initial contact with Approach Control.

After radar contact is established, you may navigate on your own into the traffic pattern or, depending on traffic conditions, you may be directed to fly specific headings to position your flight behind a preceding aircraft in the approach sequence. When your flight is positioned behind the preceding aircraft and you report having that aircraft in sight, you will be directed to follow it. If other "non-participating" or "local" aircraft are in the traffic pattern, the tower will issue a landing sequence. Upon being told to contact the tower, radar service is automatically terminated.

Standard radar separation will be provided between IFR aircraft until such time as the aircraft is sequenced and you see the traffic you are to follow. Standard radar separation between VFR or between VFR and IFR aircraft will not be provided.

If you are a departing VFR aircraft, you are encouraged to request radar traffic information by notifying ground control on initial contact with your request and proposed direction of flight. For example: "Xray Ground Control, N18 at hangar 6, ready to taxi, VFR southbound, have information Bravo and request radar traffic information." Following takeoff, the tower will advise when to contact departure control.

If you are transiting the area and in radar contact/communication with approach control, you will receive traffic information on a controller-workload-permitting basis. You should give your position, altitude, radar beacon code (if transponder equipped), destination, and/or route of flight.

Stage III Service (Radar Sequencing and Separation Service for VFR Aircraft) has been implemented at certain terminal locations. The purpose of this service is to provide separation between all participating VFR aircraft and all IFR aircraft operating within the airspace defined as the Terminal Radar Service Area (TRSA) (see Appendix B for a list of TRSAs). Pilot participation is urged but it is not mandatory. If any aircraft does not want the service, the pilot should state "Negative stage III" or make a similar comment on initial contact with Approach Control or Ground Control, as appropriate.

While operating within a TRSA, you are provided Stage III service and separation as prescribed herein. In the event of a radar outage, separation and sequencing of VFR aircraft will be suspended as this service is dependent on radar. You will be advised that the service is not available and issued wind, runway information, and the time and place to contact the tower. Traffic information will be provided on a workload-permitting basis.

Visual separation is used when prevailing conditions permit and it will be applied as follows:

- ☐ When a VFR flight is positioned behind the preceding aircraft and the pilot reports having that aircraft in sight, he will be directed to follow it. Upon being told to contact the tower, radar is automatically terminated.
- ☐ When IFR flights are being sequenced with other traffic and the pilot reports the aircraft he is to follow in sight, the pilot may be directed to follow it and will be cleared for a "visual approach."
- ☐ If other "non-participating" or "local" aircraft are in the traffic pattern, the tower will issue a landing sequence.
- ☐ Departing VFR aircraft may be asked if they can visually follow a preceding departure out of the TRSA. If the pilot concurs, he will be directed to follow it until leaving the TRSA. Until visual separation is obtained, standard vertical or radar separation will be provided; 1,000 feet vertical separation may be used between IFR aircraft; 500 feet vertical separation may be used between VFR aircraft, or between a VFR and an IFR aircraft.

- ☐ Radar separation varies depending on size of aircraft and aircraft distance from the radar antenna. The minimum separation used will be 1½ miles for most VFR aircraft under 12,500 pounds GWT. If being separated from larger aircraft, the minimum is increased appropriately.

Pilots operating VFR under Stage III in a TRSA—

- ☐ Must maintain an altitude when assigned by ATC unless the altitude assignment is to maintain at or below a specified altitude. ATC may assign altitudes for separation that do not conform to FAR 91.109. When the altitude assignment is no longer needed for separation or when leaving the TRSA, the instruction will be broadcast, "Resume appropriate VFR altitudes." Pilots must then return to an altitude that conforms to FAR 91.109 as soon as practicable.
- ☐ When not assigned an altitude, the pilot should coordinate with ATC prior to any altitude change.
- ☐ Within the TRSA, traffic information on observed but unidentified targets will, to the extent possible, be provided all IFR and participating VFR aircraft. If you request it, you will be vectored to avoid the observed traffic, insofar as possible, provided you are within the airspace under the jurisdiction of the controller.
- ☐ Departing aircraft should inform ATC of their intended destination and/or route of flight and proposed cruising altitude.
- ☐ ATC will normally advise participating VFR aircraft when leaving the geographical limits of the TRSA. Radar service is not automatically terminated with this advisory unless specifically stated by the controller.

Special note from the FAA re pilot's responsibility: *"These programs are not to be interpreted as relieving pilots of their responsibility to see and avoid other traffic operating in basic VFR weather conditions, to maintain appropriate terrain and obstruction clearance, or to remain in weather conditions equal to or better than the minima required by FAR 91.105. Whenever compliance with an assigned route, heading and/or altitude is likely to compromise said pilot responsibility respecting terrain and obstruction clearance and weather minima, Approach Control should be so advised and a revised clearance or instruction obtained.*

PRACTICE INSTRUMENT APPROACHES

ATC controls practice instrument approaches at controlled airports. Practice instrument approaches are considered by ATC to be instrument approaches made by either a VFR aircraft not on an IFR flight plan or an aircraft on an IFR flight plan. To achieve this and thereby enhance air safety, it is Air Traffic Service policy to provide for separation of such operations at locations where approach control facilities are located and, as resources permit, at certain other locations served by Air Route Traffic Control Centers or approach control facilities.

If you want to carry out practice instrument approaches, these may be approved by ATC subject to traffic and workload conditions. You should anticipate that in some instances the controller may find it necessary to deny approval or withdraw previous approval when traffic conditions warrant. It must be clearly understood, however, that even though the controller may be providing separation, if you are operating VFR, you are required to comply with basic visual flight rules (FAR 91.105).

Application of ATC procedures or any action taken by the controller to avoid traffic conflictions does not relieve IFR and VFR pilots of their responsibility to see and avoid other traffic while operating in VFR conditions (FAR 91.67). In addition to the normal IFR separation minima (which includes visual separation) during VFR conditions, 500 feet vertical separation may be applied between VFR aircraft and between a VFR aircraft and the IFR aircraft. If you are not on an IFR flight plan, and you want to do practice instrument approaches, you should always state "practice" when making requests to ATC.

Controllers will instruct VFR aircraft requesting an instrument approach to maintain VFR. This is to preclude misunderstandings between the pilot and controller as to the status of the aircraft. If you want to proceed in accordance with IFR, you must specifically request and obtain, an IFR clearance.

Before practicing an instrument approach, you should inform the approach control facility or the tower of the type of practice approach you want to make and how you intend to terminate it, i.e., full-stop landing, touch-and-go, or missed or low-approach maneuver. This information may be furnished progressively when conducting a series of approaches. If you are on an IFR flight plan, and have made a series of instrument approaches to full stop landings, you should inform ATC when you make your final landing. The controller will control flights practicing instrument approaches so as to ensure that they do not disrupt the flow of arriving and departing itinerant IFR or VFR aircraft. The priority afforded itinerant aircraft over practice instrument approaches is not intended to be so rigidly applied that it causes grossly inefficient application of services. A minimum delay to itinerant traffic may be appropriate to allow an aircraft practicing an approach to complete that approach.

At airports without a tower, if you want to make practice instrument approaches, you should notify the facility having control jurisdiction of the desired approach as indicated on the approach chart. All approach control facilities and Air Route Traffic Control Centers are required to publish a facility bulletin depicting those airports where they provide standard separation to both VFR and IFR aircraft conducting practice instrument approaches.

The controller will provide approved separation between both VFR and IFR aircraft when authorization is granted to make practice approaches to airports where an approach control facility is located and to certain other airports serviced by approach control or an ARTCC. Controller responsibility for separation of VFR aircraft begins at the point where the approach clearance becomes effective, or when the aircraft enters TRSA/TCA airspace, whichever comes first.

VFR aircraft practicing instrument approaches are not automatically authorized to execute the missed approach procedure. This authorization must be specifically requested by the pilot and approved by the controller. Separation will not be provided unless the missed approach has been approved by ATC.

Except in an emergency, aircraft cleared to practice instrument approaches must not deviate from the approved procedure until cleared to do so by the controller.

At radar approach control locations when a full approach procedure (procedure turn, etc.) cannot be approved, you should expect to be vectored to a final approach course for a practice instrument approach that is compatible with the general direction of traffic at that airport.

When granting approval for a practice instrument approach, the controller will usually ask you to report to the tower prior to or over the final approach fix inbound.

When you receive authorization to conduct practice instrument approaches to an airport with a tower, but where approved standard separation is not provided to you, the tower will approve the practice approach, instruct you to maintain VFR and issue traffic information, as required.

When you notify a Flight Service Station (FSS) by providing Airport Advisory Service of intent to conduct a practice instrument approach and if separation will be provided, you will be instructed to contact the appropriate facility on a specified frequency prior to initiating the approach. At airports where separation is not provided, the FSS will acknowledge the message and issue known traffic information but will neither approve or disapprove the approach.

When conducting practice instrument approaches, you should be particularly alert for other aircraft operating in the local traffic pattern or in proximity to the airport.

OPTION APPROACH

The "cleared for the option" procedure will permit an instructor, flight examiner, or pilot the option to make a touch-and-go, low approach, missed approach, stop-and-go, or full-stop landing. This procedure can be very beneficial in a training situation in that neither the student pilot nor examinee would know what maneuver would be accomplished. You should make your request for this procedure when passing the final approach fix inbound on an instrument approach of entering downwind for a VFR traffic pattern. The advantage of this procedure as a training aid are that it enables an instructor/examiner to obtain the reaction of a trainee/examinee under changing conditions, you would not have to discontinue an approach in the middle of the procedure due to an error or pilot proficiency requirements, and it allows more flexibility and economy in training programs. This procedure is only to be used at those locations with an operational control tower and is subject to ATC approval/disapproval.

SPECIAL VFR CLEARANCE

An ATC clearance must be obtained prior to operating within a control zone when the weather is less than that required for VFR flight. As a VFR pilot, you may request and be given a clearance to enter, leave, or operate within most control zones in special VFR conditions, traffic permitting, and providing such flight will not delay IFR operations. All special VFR flights must remain clear of clouds. The visibility requirements for Special VFR fixed-wing aircraft are one mile flight visibility for operations within the control zone and one mile ground visibility if taking off or landing. When a control tower is located within the control zone, requests for clearance should be to the tower. If no tower is located within the control zone, a clearance may be obtained from the nearest tower, FSS, or Center.

It is not necessary to file a complete flight plan with the request for clearance but you should state your intentions in sufficient detail to permit ATC to fit your flight into the traffic flow. The clearance will not contain a specific altitude as you must remain clear of clouds. The controller may require you to fly at or below a certain altitude due to other traffic, but the altitude specified will permit flight at or above the minimum safe altitude. In addition, at radar locations, flights may be vectored if necessary for control purposes or on pilot request.

Special VFR clearances are effective within control zones only. ATC does not provide separation after an aircraft leaves the control zone on a special VFR clearance.

Special VFR operations by fixed-wing aircraft are prohibited in some control zones due to the volume of IFR traffic. (See the list of airports where Special IFR operations are not permitted in Chapter 1.)

ATC provides separation between special VFR flights and between them and other IFR flights. Special VFR operations by fixed-wing aircraft are prohibited between sunset and sunrise unless the pilot is instrument rated and the aircraft is equipped for IFR flight.

TRAFFIC ADVISORIES WHEN ENROUTE

You can usually obtain radar traffic advisories when you are flying cross-country. At the lower altitudes, these will most likely come from local radar approach control units. Generally, above 7,000 feet AGL, you should ask the appropriate ATC center for service. Their frequencies are listed directly on enroute IFR charts. As always, a transponder is most helpful. If you don't have one, you probably won't be served.

PILOT/CONTROLLER ROLES AND RESPONSIBILITIES

Here are some more notes from the FAA:

The pilot in command of an aircraft is directly responsible for and is the final authority as to the safe operation of that aircraft. In an emergency requiring immediate action, the pilot in command may deviate from any rule in accordance with FAR 91.3.

The air traffic controller is responsible to give first priority to the separation of aircraft and to the issuance of radar safety advisories, second priority to other services that are required but do not involve separation of aircraft, and third priority to additional services to the extent possible.

Air Traffic Clearance

Pilot

- ☐ Acknowledge receipt and understanding of an ATC clearance. Request clarification or amendment, as appropriate, any time a clearance is not fully understood, or considered unacceptable from a safety standpoint.
- ☐ Comply with an air traffic clearance upon receipt except as necessary to cope with an emergency. If deviation is necessary, advise ATC as soon as possible and obtain an amended clearance.

Controller

- ☐ Issue appropriate clearances for the operation being or to be conducted in accordance with established criteria.
- ☐ Assign altitudes in IFR clearance that are at or above the minimum IFR altitudes in controlled airspace.

Radar Vectors

Pilot

- ☐ Comply with headings and altitudes assigned to by the controller.
- ☐ Question any assigned heading or altitude believed to be incorrect.
- ☐ If operating VFR and compliance with any radar vector or altitude would cause a violation of any FAR, advise ATC and obtain a revised clearance or instruction.

Controller

- ☐ Vector aircraft in controlled airspace:
  —For separation.
  —For noise abatement.
  —To obtain an operational advantage for the pilot or controller.
- ☐ Vector aircraft in controlled and uncontrolled airspace when requested by the pilot.
- ☐ Vector IFR aircraft at or above minimum vectoring altitude.

- ☐ May vector VFR aircraft, not at an ATC-assigned altitude, at any altitude. In these cases, terrain separation is the pilot's responsibility.

Safety Advisory

Pilot

- ☐ Initiate appropriate action if a safety advisory is received from ATC.
- ☐ Be aware that this service is not always available and that many factors affect the ability of the controller to be aware of a situation in which unsafe proximity to terrain, obstructions or another aircraft may be developing.
- ☐ This service is not a substitute for pilot adherence to safe operating practices.

Controller

- ☐ Issue a safety advisory if he is aware an aircraft under his control is at an altitude which, in the controller's judgment, places the aircraft in unsafe proximity to terrain, obstructions or another aircraft. Types of safety advisories are:
- ☐ Terrain/obstruction advisory
  Immediately issued to an aircraft under his control if he is aware the aircraft is at an altitude believed to place the aircraft in unsafe proximity to terrain/obstructions.
- ☐ Aircraft Conflict Advisory
  Immediately issued to aircraft under his control if he is aware of an aircraft not under his control at an altitude believed to place the aircraft in unsafe proximity to each other. With the alert, he offers the pilot an alternative if feasible.
- ☐ Discontinues further advisories if informed by the pilot that he is taking action to correct the situation or that he has the other aircraft in sight.

See and Avoid

Pilot

- ☐ When meteorological conditions permit, regardless of type of flight plan, whether or not under control of a radar facility, the pilot is responsible to see and avoid other traffic, terrain or obstacles.

Controller

- ☐ Provide radar traffic information to radar identified aircraft operating outside positive control airspace on a workload permitting basis.
- ☐ Issue a safety advisory to an aircraft under his control if he is aware the aircraft is at an altitude believed to place the aircraft in unsafe proximity to terrain, obstructions or other aircraft.

Traffic Advisories

Pilot

- ☐ Acknowledge receipt of traffic advisories.
- ☐ Inform controller if traffic in sight.
- ☐ Advise ATC if a vector to avoid traffic is desired.
- ☐ Do not expect to receive radar traffic advisories on all traffic. Some aircraft may not appear on the radar display. Be aware that the controller may be occupied with higher priority duties and unable to issue traffic information for a variety of reasons. Advise controller if service is not desired.

Controller

- ☐ Issue radar traffic to the maximum extent consistent with higher priority duties except in positive controlled airspace. Provide vectors to assist aircraft to avoid observed traffic when requested by the pilot.
- ☐ Issue traffic information to aircraft in the airport traffic area for sequencing purposes.

Visual Separation

Pilot

- ☐ Acceptance of instructions to follow another aircraft or to provide visual separation from it is an acknowledgment that you will maneuver your aircraft as necessary to avoid the other aircraft or to maintain in-trail separation.

- ☐ If instructed by ATC to follow another aircraft to provide visual separation from it, promptly notify the controller if you lose sight of that aircraft, are unable to maintain continued visual contact with it, or cannot accept the responsibility for your own separation for any reason.
- ☐ The pilot also accepts responsibility for wake turbulence separation under these conditions.

Controller

- ☐ Apply visual separation only within a terminal area when a controller has both aircraft in sight or by instructing a pilot who sees the other aircraft to maintain visual separation from it.

COLLISION AVOIDANCE SUMMARY

Collision avoidance is a combination of:

- ☐ Knowing and observing FARs.
- ☐ Understanding the airspace structure.
- ☐ Keeping a sharp lookout for traffic at all times, and using good scan techniques.
- ☐ Making allowances for your airplane's visibility shortcomings.
- ☐ Using radar services where possible.
- ☐ Staying within your legal and operational limitations.

COLLISION AVOIDANCE QUIZ

Answers are on page 181.

1. You are approaching an unfamiliar uncontrolled airport. There appears to be no Unicom at the field. What should you do?

 a. Make blind position reports on 122.7.
 b. Make blind position reports on 122.8.
 c. Make blind position reports on 122.9.
 d. Make blind position reports on 123.0.

2. An airport has no tower, but does have an operational FSS. What air traffic communications can you expect there?

a. Airport Advisory Service on 123.0.
b. Airport Advisory Service on a published frequency.
c. Air traffic control service on Airport Advisory Service frequency.
d. Wind and runway information only.

3. IFR charts are reissued every:

a. 56 days.
b. 60 days.
c. 90 days.
d. 6 months.

4. Most Sectional charts are reissued:

a. Every 90 days.
b. Every 6 months.
c. Every year.
d. As necessary.

5. You are receiving traffic service from a radar approach control facility. You are operating VFR. The controller instructs you to "Turn right to 090° *immediately* for traffic avoidance." This heading will put you into a cloud, which is based about 500 feet below you. What do you do?

a. Turn right to 090° and advise that you will enter a cloud on that heading.
b. Keep going straight ahead and advise that you would enter a cloud on the 090° heading.
c. Start turning right immediately, advise that the 090° heading would put you into a cloud, and suggest that 120° would be better.
d. Make an immediate diving turn to the right onto 090°, so as to avoid the cloud.

6. Stage I TRSA is:

a. Radar advisory service for VFR aircraft.

b. Radar advisory and sequencing for VFR aircraft.
c. Radar sequencing and separation for VFR aircraft.
d. Radar vectoring and traffic service for VFR aircraft.

7. Stage II TRSA is:
a. Radar advisory service for VFR aircraft.
b. Radar advisory and sequencing for VFR aircraft.
c. Radar sequencing and separation for VFR aircraft.
d. Radar vectoring and traffic service for VFR aircraft.

8. Stage III TRSA is:
a. Radar advisory service for VFR aircraft.
b. Radar advisory and sequencing for VFR aircraft.
c. Radar sequencing and separation for VFR aircraft.
d. Radar vectoring and traffic service for VFR aircraft.

9. You want to carry out a VFR practice ILS approach at a controlled airport. To be able to initiate this, you must:
a. File a VFR flight plan.
b. File a IFR flight plan.
c. Contact the tower of the airport.
d. Contact the approach control of the airport.

10. The responsibility for traffic, terrain and obstacle avoidance in VFR conditions while being radar vectored is that of:
a. The pilot in command.
b. Air Traffic Control.
c. Both the pilot in command and ATC.
d. If an instrument check flight, the examiner.

Chapter 5
Wake Turbulence

Every airplane creates a wake as it flies through the air. For a time, pilots called this wake "propwash." However, t'ain't so. The wake is caused by a pair of counterrotating vortices trailing from the wing. The *larger* the aircraft, the more extensive the vortices. And the *slower* the aircraft, the more extensive the vortices. And the *cleaner* the aircraft (e.g. flaps and gear up), the more extensive the vortices. In tests, peak velocities at the edge of the vortex on a large aircraft were recorded at about 133 knots.

A very large, slow, clean aircraft can impose a rolling moment that exceeds the roll-control capability of some airplanes. This means that if you encounter such a vortex, you *cannot* overcome the resultant upset to your flight by movement of your controls. Further, turbulence generated within the vortices can damage parts of your aircraft or your equipment if you hit it close behind a big one. You must learn to visualize the location of the vortex wake generated by large aircraft and adjust your flight path accordingly.

HOW VORTICES ARE GENERATED

Lift is created by the difference in pressure between the top and bottom surfaces of the wing. The top surface has lower pressure than the bottom surface. This pressure

Large, slow and clean, the Lockhead C25A Galaxy is one to avoid! (courtesy USAF)

differential is settled at the wing tips, causing the two airflows where they meet to roll in a swirling air mass trailing rearwards from the tips. Thus the wake consists of two counterrotating cylindrical vortices. Vortex characteristics of any given aircraft can be changed by variations in speed, as well as extension of flaps or slats.

INDUCED ROLL

It has happened that a wake encounter has resulted in structural damage of catastrophic proportions. In flight experiments, aircraft have been intentionally flown directly up trailing vortex cores of large aircraft. It was found that an aircraft's ability to counteract a vortex-induced roll depends chiefly on its wingspan and control responsiveness. The ideal situation is found where wingspan and ailerons extend beyond the rotational flow field of the vortex. Aircraft with short wingspans have greater difficulty in dealing with the roll caused by vortex flow. Large aircraft wakes demand your respect!

VORTEX BEHAVIOR

You can visualize wake location and thus take evasive action if you remember the following points:

When you and a large aircraft are sharing a runway or using adjacent runways, remember that vortices are generated from the moment the aircraft leaves the ground, since wing lift creates them.

Before takeoff or touchdown, note the rotation or touchdown point of the preceding aircraft. Avoid the wake by maintaining a flight path *above* the vortices.

Remember that wind blows vortices as a mass, so, for example, if you have been cleared for takeoff behind a DC-10, and there is a good wind from the right, allow enough time for the wake to drift to the left, and stay to the right side of the runway. Large airport runways are wide enough that a light airplane pilot need not line up on the centerline. If necessary, line up so that your wingtip is over the grass, and as soon as you are airborne, make an adjustment to the upwind side.

If the wind is down the runway, or the wind is calm, it is safer to wait until the vortex has dissipated. Even with the cleaner burning engines of today, you can still see some exhaust smoke from most jets. This smoke trail can be a good guide as to what is happening to the wake. Re-

Keep clear of Concorde on takeoff. (courtesy British Airways)

member, however, that the wake will be *outside* of the smoke trial. Watch the smoke trail, and guide your airplane to avoid it, making plenty of allowance beyond the edge of the smoke.

If you are using a runway parallel to one that was used to depart a heavy aircraft, remember that a crosswind coming from the other runway could blow the wake across your runway just as you are taking off or landing. If possible, where parallel runways exist and there is a crosswind, try to use the upwind runway, especially if traffic is heavy and the runways are close together.

The vortex circulation is outward, upward and around the wingtips when viewed from either ahead or behind the aircraft.

Tests with large aircraft have shown that the vortex flow field, in a plane cutting through the wake at any point downstream, covers an area about two wingspans in width and one span in depth. The vortices remain so spaced (about a wingspan apart), drifting with the wind, at altitudes greater than a wingspan from the ground. In view of this, if

Boeing 747 landing at JFK. (courtesy Port Authority of NY & NJ)

you encounter persistent vortex turbulence, make a slight change of altitude (climbing) and lateral position (upwind).

Vortices from large aircraft sink at a rate of about 400 to 500 feet per minute. They tend to level off at a distance about 900 feet below the flight path of the generating aircraft. Vortex strength diminishes with time and distance behind the generating aircraft. Atmospheric turbulence hastens breakup.

When taking off from an intersection, thus using only part of the runway, be particularly aware of wake hazards, since you are more likely to be close to the rotation point of a previously departing aircraft.

Think hard about what is going on when operating from a large airport, and adjust your thinking to reflect reality. I had an experience once I shall never forget in which the events were abnormal, and I had failed to recognize the difference they caused. Here's what happened: I was waiting for takeoff from runway 28 at Toronto International Airport. I was flying a Tri-Pacer, taking some friends for a joyride (this was many years ago—1958). Ahead of me, also awaiting clearance was another Tri-Pacer and a DH Dove. An Air Canada Viscount was on final approach, less than a mile out. Suddenly, and without clearance, the Tri-Pacer ahead pulled out onto the runway and started to take off. At this moment the Viscount was over the approach lights. The tower immediately told the Viscount to go around, and he pulled up, screaming down the runway at about 100 feet. He passed a few feet over the sinful Tri-Pacer, doubtless giving him an unforgettable experience! The tower then cleared the Dove to take off, and, maybe 30 seconds later, cleared me out. I took off, and started climbing. At about 100 feet, the airplane was suddenly wrested from my control and put on its right side at a 90° bank angle. I immediately applied full opposite everything, and the Tri-Pacer recovered. We continued with an uneventful flight, interrupted only by the racing of four hearts. What happened? I had flown right into the Viscount's wake—yet he had done his overshoot a good minute beforehand. I had not thought about the wake hazard, because I was quite inexperienced then (I only had 113 hours at the time), and I was taking off behind a little Dove, not a Viscount. I failed to recognize the Viscount overshoot as a hazard. We live and learn, as they say.

When landing behind a big aircraft that is also landing, fly at or above its flight path, altering course as necessary to avoid the area behind and below it.

When landing on a runway from which a big aircraft has just taken off, land short, so as to be on the ground before his liftoff point. But remember, a headwind will blow the wake toward you, shortening the amount of safe runway available. So if necessary, wait until the wake area has passed by slowing down or S-turning, go around, or refuse the runway and ask for another one. You don't *have* to accept a runway you have been assigned. You are the captain, not the tower controller.

A crossing runway can be equally hazardous, since you may fly your aircraft through the wake of another aircraft on the other runway. Plan your flight to be above the other aircraft's wake.

Keep clear of Concorde on landing, too. (courtesy British Airways)

VORTEX MOVEMENT

When the vortices of large aircraft sink close to the ground (within about 200 feet), they tend to move laterally over the ground at a speed of about 5 knots. A crosswind will decrease the lateral movement of the upwind vortex and increase the movement of the downwind vortex. Thus a light wind of 3 to 7 knots could result in the upwind vortex remaining in the touchdown zone for a period of time and hasten the drift of the downwind vortex toward another runway. Similarly, a tailwind condition can move the vortices of the preceding aircraft forward into the touchdown zone. A light quartering tailwind requires maximum caution.

A wake encounter is not necessarily hazardous. It can be one or more jolts with varying severity depending upon the direction of the encounter, distance from the generating aircraft, and point of vortex encounter. The probability of induced roll increases when the encountering aircraft's heading is generally aligned with the vortex trail or flight path of the generating aircraft. Always avoid the area below and behind large aircraft, especially at low altitude where even a momentary wake encounter could be hazardous.

Jet blast can cause damage and upsets if you encounter it at close range. Studies have been made of exhaust velocity versus distance, and have shown a need for light aircraft to keep an adequate separation during ground operations. Here is a table of sample distance requirements to avoid exhaust velocities of greater than 25 mph:

25 mph Velocity	B-727	DC-8	DC-10
Takeoff Thrust	550 ft.	700 ft.	2100 ft.
Breakaway Thrust	200 ft.	400 ft.	850 ft.
Idle Thrust	150 ft.	100 ft.	350 ft.

Engine exhaust velocities generated by large jet aircraft during initial takeoff roll and the drifting of the turbulence in relation to the crosswind component mean light aircraft awaiting takeoff should hold well back of the runway edge of the taxiway hold line; also, it is a good idea to align the aircraft to *head into* the jet blast. Additionally, in the course of running up engines and taxiing on the ground, large aircraft can produce powerful jet blasts that can flip a light aircraft over.

IN FLIGHT

Always avoid flight below and behind a large aircraft's path. If you see you will cross behind another aircraft, try to be above where his wake would fall. When flying in a TCA, you will often find you have to maintain an assigned VFR altitude (which will usually be a thousand-foot altitude plus 500 feet), and may find large aircraft operating 500 feet above you. You may not be able to change altitude in this case. And you may be flying an assigned heading, with no deviation allowed. What do you do? Tell the controller you want to turn xxx degrees to avoid wake turbulence, and get out of the way. If necessary, do a 180. If you can't get a word in edgewise, due to a lot of talk on the frequency, take the evasive action anyway and tell the controller as soon as you can attract his attention. Your safety is important and you are in command, and thus responsible for avoiding a hazardous situation.

HELICOPTERS

Hovering helicopters generate a downwash from their main rotors similar to the prop blast of a conventional aircraft. However, in forward flight, this energy is transformed into a pair of trailing vortices similar to wingtip vortices of fixed-wing aircraft. You should avoid the vortices as well as the downwash.

SUMMARY

Wingtip vortices and jet blast on the ground can be very dangerous, with results as catastrophic as a midair collision. Be as wary of these as you are of actually hitting another airplane.

WAKE TURBULENCE QUIZ

Answers are on page 181.

1. You have been cleared to land following a DC 10.

Bell 222. (courtesy Bell)

Correct wake-turbulence avoidance technique involves:

a. Staying at least three miles behind the heavy aircraft.
b. Remaining well below the flight path at all times.
c. Remaining well above his flight path at all times.
d. None of the above.

2. You have been cleared to take off behind a Boeing 747 on runway 32. The wind is from the north at 30 knots. You should:

a. Wait at least three minutes before taking off.
b. Line up on the right side of the runway and turn slightly right when in the air.
c. Line up on the left side of the runway and turn slightly left when in the air.
d. Ask for another runway.

3. Wake turbulence from a heavy aircraft is at its worst when the heavy aircraft is:

a. Fast and clean (gear/flaps up).
b. Slow and clean (gear/flaps down).
c. Fast and dirty (gear/flaps down).

d. Slow and dirty (gear/flaps down).

4. You are enroute and you notice you will pass within a half a mile behind a 747 which is climbing out to its cruising altitude. You should:

a. Cross above the flight path of the 747 at 90°.
b. Cross below the flight path of the 747 at 90°.
c. Do a 360°.
d. Turn and fly parallel to the flight path for three minutes, while staying above.

5. You are holding in your light aircraft for takeoff clearance at Kennedy airport, in a line of heavy aircraft. A new takeoff clearance is issued every two minutes, and each time, the line moves forward one slot. How far behind the Lockheed 1011 that is in front of you should you wait?

a. 2,100 feet
b. 850 feet
c. 350 feet
d. 150 feet

Chapter 6
Transponder Operation

Proper use of your transponder will provide you with a higher degree of safety in the environment where high-speed closure rates are possible. Transponders substantially increase the capability of radar to see an aircraft and the altitude reporting (mode C) feature enables the controller to quickly determine where potential traffic conflicts may exist. Regardless, never relax your visual scanning vigilance for other aircraft.

Your transponder should be turned on, or set in the "normal" operating position as late as practicable prior to takeoff and to "off" or "standby" as soon as practicable after completing your landing roll unless the change to "standby" has been accomplished previously at the request of ATC. In all cases, whether VFR or IFR, the transponder should be operating while airborne unless otherwise requested by ATC.

If you are IFR and you cancel your IFR flight plan prior to reaching your destination, you should adjust your transponder according to VFR operations (squawk 1200, or as advised by ATC).

TRANSPONDER RANGE

Bear in mind that the coverage you can expect is limited to line-of-sight. Low altitude or antenna shielding

Collins TDR 950 transponder. (courtesy Collins)

by the aircraft itself may result in reduced range. Range can be improved by climbing to a higher altitude. Your transponder antenna should be located where dead spots caused by the aircraft are only noticed during abnormal flight attitudes.

TRANSPONDER CODE DESIGNATION

When ATC uses one or a combination of the 4096 discrete transponder codes, four-digit code designation is used in communications, e.g., code 2105 is expressed as Two One Zero Five.

AUTOMATIC ALTITUDE REPORTING (MODE C)

Many transponders are equipped with a mode C automatic altitude reporting capability. This system converts aircraft altitude, in 100 foot increments, to coded digital information that is transmitted together with mode C framing pulses to the interrogating radar facility. Some altitude encoders are built into the airplane's altimeter; others are separate units. Some of the newer transponders give a display of the altitude that is being sent out so that you can cross-check it with what you see on your altimeter.

You should always use your altitude encoder, unless ATC tells you otherwise. An instruction by ATC to "Stop altitude squawk, altitude differs (*number of feet*)" may be an indication that your transponder is transmitting incorrect altitude information or that you have an incorrect altimeter setting. While an incorrect altimeter setting has no effect on the mode C altitude information transmitted by your transponder (transponders are preset at 29.92), it would cause you to fly at an actual altitude different from your assigned altitude. When a controller indicates that an altitude readout is invalid, you should initiate a check to verify that the aircraft altimeter is set correctly.

If you have an encoder, report your exact altitude or flight level to the nearest hundred-foot increment when establishing initial contact with an ATC facility. These reports on initial contact provide ATC with information that is required prior to using mode C altitude information for separation purposes. It also significantly reduces requests for verification of altitude.

TRANSPONDER "IDENT" FEATURE

You should only activate the "ident" feature when ATC asks you to do so. Doing so causes your own target to bloom on the radar screen for a few seconds, and is used by the controller to verify your position.

MAKING CODE CHANGES

When making routine code changes, avoid inadvertently selecting codes starting with 75—, 76— or 77—, which could cause momentary false alarms at automated ground facilities. For example, when switching from code 2700 to code 7200, switch first to 2200 then to 7200, *not* to 7700 and then 7200. This procedure applies to code 7500 and all discrete codes in the 7600 and 7700 series (i.e., 7600-7677, 7700-7777) which will trigger special indicators in automated facilities.

Under no circumstances should you operate the transponder on code 0000. This code is reserved for military interceptor operations.

King RT79 transponder shows the altitude it is sending to ATC. (courtesy King)

DEVIATIONS FROM TRANSPONDER REQUIREMENTS

Specific details concerning requirements, exceptions, and ATC authorized deviations for transponder and mode C operation above 12,500 feet and below 18,000 feet MSL are found in FAR 91.24. In general, the FAR requires aircraft to be equipped with mode A (4096 codes) and mode C altitude reporting capability when operating in controlled airspace of the 48 contiguous states and the District of Columbia above 12,500 feet MSL, excluding airspace at and below 2,500 feet AGL. You should ensure that your transponder is operating on an appropriate or ATC-assigned VFR/IFR code and mode C when operating in such airspace. If in doubt about the operational status of either feature of your transponder while airborne, contact the nearest ATC facility or FSS to find out what facility you should contact to check your equipment.

If your transponder fails in flight and you are in airspace where a transponder is required, you can make a request for immediate deviation from transponder requirements, and this may be approved by controllers only when the flight will continue IFR or when weather conditions prevent VFR descent and continued VFR flight in airspace not affected by the FAR.

All other requests or deviation should be made by contacting the nearest FSS or ATC facility in person or by telephone. The nearest ARTC Center will normally be the controlling agency and is responsible for coordinating requests involving deviations in other ARTCC areas.

TRANSPONDER OPERATION WHEN VFR

Unless otherwise instructed by an ATC facility, set your transponder to 1200, regardless of altitude, when operating VFR. If you have an altitude encoder, use this feature, unless deactivation is directed by ATC or unless the installed equipment has not been tested and calibrated as required by FAR 91.36.

TRANSPONDER OPERATION DURING EMERGENCY CONDITIONS

If an emergency occurs and you can't establish immediate communications with a ground radar facility, you can alert them by adjusting your transponder to reply on code 7700. You may not be within a radar coverage area and even if you are, some radar facilities are not yet equipped to automatically recognize code 7700 as an emergency signal. Therefore, you should establish radio communications with an ATC facility as soon as possible.

TRANSPONDER OPERATION DURING RADIO FAILURE

If you experience a loss of two-way radio capability, you should: Adjust your transponder to reply on code 7700 for a period of one minute, then change to code 7600 and remain on 7600 for a period of 15 minutes or the remainder of the flight, whichever occurs first. Repeat these steps as practicable.

TRANSPONDER PHRASEOLOGY IN COMMUNICATIONS

Air traffic controllers, both civil and military, use the following phraseology when referring to operation of your transponder (also known as the Air Traffic Control Radar Beacon System (ATCRBS).

squawk (number)—Operate radar beacon transponder on designated code in mode A.

ident—Engage the "ident" feature of the transponder.

squawk (number) and ident—Operate transponder on specified code in and engage the "ident" feature.

squawk standby—Switch transponder to standby position.

squawk low-normal—Operate transponder on low or normal sensitivity as specified. Transponder is operated in "NORMAL" position unless ATC specifies "LOW" ("ON" is used instead of "NORMAL" as a master control label on some types of transponders.)

squawk altitude—Activate mode C with automatic altitude reporting.

stop altitude squawk—Turn off altitude reporting switch and continue transmitting mode C framing pulses. If your equipment does not have this capability, turn off mode C.

stop squawk—Switch off transponder.

squawk mayday code 7700—Operate transponder in the emergency position (mode A code 7700).

SUMMARY

A transponder is essential for effective flight operations in high-density traffic areas, and an encoding altimeter is strongly recommended. Proper understanding and use your transponder will ease your trips and reduce confusion. Don't even think of flying in congested airspace without one!

TRANSPONDER QUIZ

Answers are on page 182.

1. You are enroute VFR, talking to no one, enjoying the scenery. You should keep your transponder:

a. On 1200.
b. On 1200 with altitude reporting, if available.
c. Off.
d. 0400.

2. When taking off, you should turn your transponder on:

a. As you start your takeoff roll.
b. After contacting Departure Control.
c. Before taxiing out.
d. When instructed to by ATC.

3. The emergency transponder code is:

a. 0700.
b. 1200.
c. 7600.
d. 7700.

4. The radio-failure transponder code is:

a. 7600.
b. 1200.
c. 7700 for 1 minute, then 7600 for 15 minutes, repeated.
d. 7700.

5. Squawking "ident" causes your radar target to:
a. Bloom on the radar scope for 5 seconds.
b. Bloom on the radar scope.
c. Flash.
d. Read out a data block.

Chapter 7
Flying into Major Airports

Some big airports absolutely hate us little guys, and prove it by charging outrageous landing/takeoff fees. The two worst are New York's La Guardia and Kennedy, at which during peak hours, you can expect to get hit with an operating fee of $50 to $70 and more.

For example, Kennedy has peak hours of 1500 to 2200 local (3 p.m. to 10 p.m.), seven days a week. During these times, the operating fee is 80¢ per 1,000 pounds gross takeoff weight of your aircraft, plus a $50 surcharge. So a 3,000 pound airplane would cost $52.40 to land or take off in peak hours. Their off-peak rate has a minimum of $20.

At La Guardia, peak hours are 0800 to 2100 (8 a.m. to 9 p.m.), and their rate is $2.50 per 1,000 pounds, with a $20 minimum, *plus* a $50 surcharge for peak hour operations.

On the other hand, Newark has a fee of a minimum of $20, with no peak surcharges. Atlanta Hatsfield charges 58¢ per 1,000 pounds with a $4 minimum. Washington National charges a $4 minimum landing fee, and Teterboro New Jersey has a $2.50 charge. Los Angeles International charges *no* landing fee (!) for non-revenue flights, and a $10 minimum for light aircraft on a revenue flight. Chicago O'Hare wants $25, and Chicago Midway only $2.50. Boston used to charge $45, and now wants only $5. So there are no

New York's Kennedy airport from above the TCA. (courtesy Port Authority of NY & NJ)

rules for rates. They vary by airport, aircraft weight, time of day and type of operation.

Apart from the costs of landing and taking off, you'll find that at many of these airline airports, the services

La Guardia—the most expensive airport in America. (courtesy Port Authority of NY & NJ)

Newark airport in 1972, while the new terminals were still under construction. Can you spot the general aviation area? (courtesy Port Authority of NY & NJ)

Boston Logan is right next to downtown. Can you spot the general aviation area? (courtesy Massport)

available for general aviation are either minimal or priced as for Boeing 747s. At the time of writing, 100-octane fuel at La Guardia is $2.10 a gallon, and at Atlanta $2.19, compared with a national average of $1.95, and a low of $1.60, according to the monthly *Professional Pilot* magazine survey. A battery jump-start at one of these airports was quoted at $50. Many of them have limited parking and tiedown facilities for light airplanes, and no maintenance services, so really the only reason to fly into one of these airports is to make a connection with an airline, or if the person you want to see is right there. Otherwise, you're much better off to go to a satellite field.

CONNECTING WITH AN AIRLINE

When you go into a major airline field to connect with a flight (say to drop off or pick up a passenger), you'll find you won't be able to park at the regular airline gates, except in very rare instances. And if you do, you'll have to go through a security check before getting into the terminal. More often, you'll have to go to the general aviation facility, probably all the way around the other side of the field, and get a ride to the terminal. Most facilities make no charge for the ride, however, although I've never seen a driver turn down a tip.

This is an airport? Yes . . . see the runway at top left? (courtesy Atlanta Hartsfield Airport)

Newark's terminals, with Manhattan in the background. (courtesy Port Authority of NY & NJ)

WHAT TO WATCH OUT FOR

So you're a glutton for punishment, and you decide to fly into an O'Hare or a Kennedy. What must you watch out for? First of all, make sure you have a good map of the taxiways that shows their numbering system. Jeppesen provides excellent ones with their Airway Manual service. Expect to do a lot of taxiing. It's not unusual to have to taxi two or three miles from where you landed to the GA terminal—in fact, it's best to plan your landing to be as close to the GA terminal as possible to reduce this problem. Some big airports have special light aircraft or STOL runways—often converted from taxiways.

Watch out for jet blast. A large jet at breakaway thrust (the power needed to get it moving) can flip you over if you're too close. See Table on page 97. Watch out for vehicles. You have the right of way over vehicles, but there are so many of them scurrying around that you must keep your eyes open. At many airports there is a vast expanse of concrete, and roads and taxiways are marked out by painted lines. Make sure you taxi on a taxiway and not a road, otherwise you may find yourself tangling with a bus or a fuel truck.

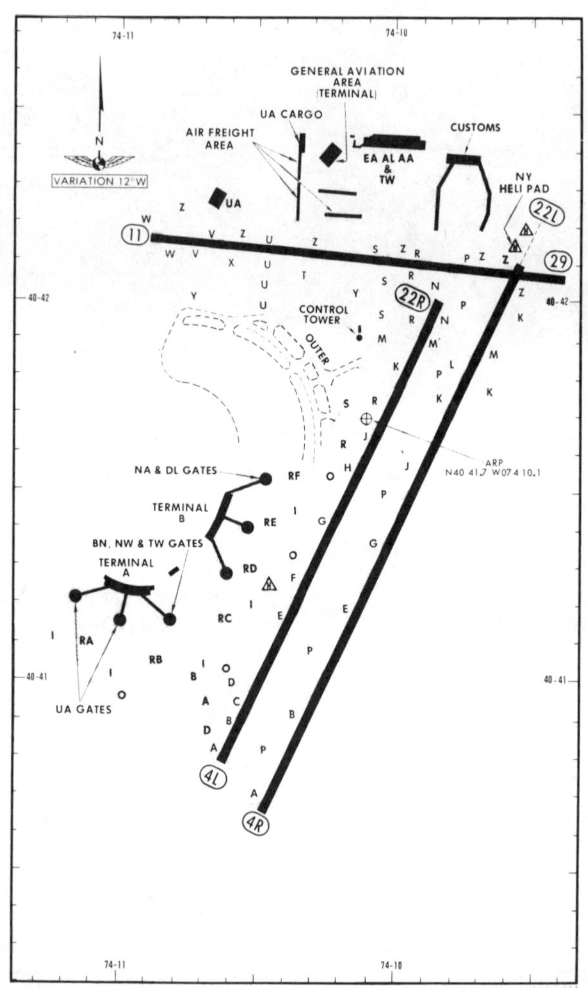

Newark Airport chart. © 1983 Jeppesen Sanderson. Not for navigation.

If you are in doubt as you meander around the airport, ask for directions from Ground Control. Bear in mind they may not even be able to see you if you're blocked by a hangar or a jumbo jet. Now, let's take a look at getting into one of these airports.

APPROACHING THE AIRPORT

Most major airports have an Approach Control, which is usually a radar facility, although there are still a few without radar. Approach is responsible for controlling all instrument flights operating within its area of responsibility, and also provides service to VFR flights. Approach Control may serve one or more airfields, and control is exercised primarily by direct pilot/controller communications. If you are VFR, first listen to the ATIS, then contact Approach Control (the frequency is on airport IFR charts and area charts). If you're IFR, the Center will tell you to contact Approach Control on a specified frequency; listen to the ATIS before you do. This is not mandatory but is strongly recommended.

Where radar is used for Approach Control service, it is used not only for traffic separation, but for radar approaches and vectors in conjunction with published non-radar approaches based on radio navaids (ILS, VOR, NDB). Radar vectors can provide course guidance and expedite traffic to the final approach course of any established instrument approach procedure or to the traffic pattern for a visual approach.

Approach Control facilities that provide radar service operate like this. Arriving IFR aircraft are cleared to an outer fix most appropriate to the route being flown with vertical separation and, if required, given holding information. If radar handoffs are effected between the Center and Approach Control, or between two Approach Control facilities, aircraft are cleared to the airport or to a fix that will enable the handoff to be completed prior to the time the aircraft reaches the fix. When radar handoffs are used, successive arriving flights may be handed off to approach control with radar separation instead of vertical separation. After release to Approach Control, aircraft are vectored to the appropriate final approach course. Headings given are required for spacing and separating aircraft. *Therefore, you must not deviate from the headings issued by Approach Control.* If you have to be vectored *across* the final IFR approach course for spacing or other reasons, you should be so advised. If it becomes apparent that your vector will take you across the final approach course, and you have not been informed that you'll be so vectored, you should query the controller. You are not expected to turn inbound on the final approach course unless an approach clearance has been issued. This clearance will normally be issued with the final vector for interception of the final approach fix. In the case of aircraft already inbound on the final approach course, approach clearance will be issued prior to the aircraft reaching the final approach fix.

When established inbound on the final approach course, radar separation will be maintained and you'll be expected to complete the approach using the approach aid designated in the clearance (ILS, VOR, radar beacons, etc.) as the primary means of navigation. Therefore, once established on the final approach course, you must not deviate from it unless a clearance to do so is received from ATC. After passing the final approach fix, you are expected to continue inbound and complete the published approach or effect the missed approach procedure published for that airport.

Whether you are vectored to the appropriate final approach course or provide your own navigation on published routes to it, radar service is automatically terminated when the landing is completed or the tower controller has you in sight, whichever comes first.

VFR traffic may be simply provided vectors to the airport, or just traffic advisories.

EXITING THE RUNWAY AFTER LANDING

After landing, unless otherwise instructed by the control tower, continue to taxi in the landing direction, proceed to the nearest suitable taxiway, and exit the runway without delay. Do not turn on to another runway or make a 180 degree turn to taxi back on an active runway or change to Ground Control while on the active runway without authorization from the tower.

The majority of Ground Control frequencies are in the 121.6—121.9 mHz bandwidth. Ground Control frequencies are provided to eliminate frequency congestion on the tower frequency and are limited to communications be-

The author landing his Comanche at Trenton, New Jersey. (courtesy Mary Foster)

tween the tower and aircraft on the ground and between the tower and utility vehicles on the airport. They are used for issuance of taxi information, clearances, and other necessary contacts between the tower and aircraft or other vehicles operated on the airport. Normally, only one Ground Control frequency is assigned at an airport; however, at locations where the amount of traffic so warrants, a second Ground Control frequency and/or another frequency designated as a Clearance Delivery frequency may be assigned.

In communications with you, a controller may omit the Ground or Local Control frequency if the controller believes you know which frequency is in use. If the ground control frequency is in the 121 mHz bandwidth the controller may omit the numbers preceding the decimal point (e.g., 121.7, ''Contact Ground point seven''). However, if any doubt exists as to what frequency is in use, ask!

Approval must be obtained prior to moving an aircraft or vehicle onto the movement area during the hours an airport traffic control tower is in operation. The movement area is normally described in local bulletins issued by the airport manager or control tower. These bulletins may be found in FSSs, FBO offices, and operations offices. The control tower also issues bulletins describing areas where they cannot provide airport traffic control service due to nonvisibility or other reasons.

A clearance must also be obtained prior to taxiing on a runway. In the absence of holding instructions, a clearance to "taxi to" any point other than an assigned takeoff runway is a clearance to cross all runways that intersect the taxi route to that point.

PROCEDURES ON DEPARTURE

When you are planning to depart the airport, first listen to the ATIS report. Then communicate with the control tower on the appropriate Ground Control/Clearance Delivery frequency prior to starting engines to receive engine start time, taxi and/or clearance information. Unless otherwise advised by the tower, remain on Ground Control frequency during taxiing and runup, then change to the tower control frequency when ready to request takeoff clearance.

TAXIING

Always state your position on the airport when calling the tower for taxi instructions. A clearance must be obtained prior to taxiing on a runway, taking off, or landing during the hours an airport traffic control tower is in operation. When ATC clears an aircraft to "taxi to" an assigned takeoff runway, the absence of holding instructions authorizes the aircraft to "cross" all runways which the taxi route intersects except the assigned takeoff runway. It does not include authorization to "taxi onto" or "cross" the assigned takeoff runway at any point.

ATC clearances or instructions pertaining to taxiing are predicated on known traffic and known physical airport conditions. Therefore, it is important that you clearly understand the clearance or instructions. Although an ATC clearance is issued for taxiing purposes, it is your responsibility as pilot in command to avoid collision with other

Mixing it up with the big guys at JFK. (courtesy Port Authority of NY & NJ)

aircraft. You should obtain clarification of any clearance or instruction you don't understand.

At those airports where the U.S. Government operates the control tower and ATC has authorized noncompliance with the requirement for two-way radio communications while operating within the airport traffic area, or at those airports where the control tower is privately operated and radio communications cannot be established, obtain a clearance by visual light signal prior to taxiing on a runway and prior to takeoff and landing. See Table 7-1 for the light signals, in case you have forgotten from your ground school days.

LIGHT SIGNALS

FAR 91.87 gives the ATC light signals as shown in Table 7-1.

TAXI DURING LOW VISIBILITY

During certain low-visibility conditions, the movement of aircraft and vehicles on airports may not be visible

Table 7-1. Light Signals.

COLOR AND TYPE OF SIGNAL	MEANING FOR AIRCRAFT ON THE SURFACE	MEANING FOR AIRCRAFT IN THE AIR
Steady green	Cleared for takeoff	Cleared to land
Flashing green	Cleared to taxi	Return for landing (followed by steady green at appropriate time)
Steady red	Stop	Give way to other aircraft and continue circling
Flashing red	Taxi clear of runway in use	Airport unsafe — do not land
Flashing white	Return to starting point on airport	Not applicable
Alternating red and green	Exercise extreme caution	Exercise extreme caution

to the tower controller. This may prevent visual confirmation of an aircraft's adherence to taxi instructions. Therefore, exercise extreme vigilance and proceed cautiously under such conditions. Of vital importance is the need to notify the controller when difficulties are encountered or at the first indication of becoming disoriented. Proceed with extreme caution when taxiing toward the sun. When vision difficulties are encountered, immediately inform the controller.

DEPARTURE CONTROL

Departure Control is responsible for ensuring separation between departures. So as to expedite their handling, Departure Control may suggest a takeoff direction other than that which may normally have been used under VFR handling. Many times it is preferred to offer the pilot a runway that will require the fewest turns after takeoff to place the pilot on his filed course or selected departure route as quickly as possible. At many locations, particular attention is paid to the use of preferential runways for local noise abatement programs and route departures away from congested areas.

Departure Control using radar will normally clear aircraft out of the terminal area using standard instrument departures via radio navigation aids. When a departure is to be vectored immediately following takeoff, you'll be advised prior to takeoff of the initial heading to be flown but may not be advised of the purpose of the heading. Pilots operating in a radar environment are expected to associate departure headings with vectors to their planned route or flight. When given a vector taking you off a previously assigned nonradar route, you'll be advised briefly what the vector is to achieve. Thereafter, radar service will be provided until the aircraft has been re-established "on-course" using an appropriate navigation aid and the pilot has been advised of this position, or a handoff is made to another radar controller with further surveillance capabilities.

Controllers will tell you the Departure Control frequencies and, if appropriate, the transponder code before takeoff. Don't operate your transponder until ready to start the takeoff roll, and don't change to the Departure Control frequency until requested.

WHICH AIRPORT IS BEST?

As I've indicated, many airports don't make you feel welcome if you're little. So where do you go when you want to fly to one of the major cities? Herewith, a list of suggestions, based on Group I TCAs. However, some of these alternates are very busy with general aviation traffic. See Appendix A.

Group I Terminal Control Areas

State	*City*	*Main Airport to Avoid*	*Suggested General Aviation Alternates*
CA	Los Angeles	LAX International	BUR Burbank
			HHR Hawthorne
			LGB Long Beach
			SMO Santa Monica
			VNY Van Nuys
CA	San Francisco	SFO International	HWD Hayward
			OAK Oakland
DC	Washington	DCA National	BWI Baltimore-Washington International
			CGS College Park MD
			IAD Dulles VA
			GAI Gaithersburg MD
FL	Miami	MIA International	FLL Fort Lauderdale
			FXE Fort Lauderdale
			HWO Hollywood
			OPF Opa Locka
			TMB Tamiami
GA	Atlanta	ATL Hartsfield	FTY Fulton County
			PDK DeKalb Peachtree
IL	Chicago	ORD O'Hare	DPA Du Page

State	City	Main Airport to Avoid	Suggested General Aviation Alternates	
			CGX	Meigs
			MDW	Midway
			PWK	Pal-Waukee
MA	Boston	BOS Logan	BED	Hanscom (Bedford)
			OWD	Norwood
NY	New York	LGA La Guardia	CDW	Essex City NJ (Caldwell)
		JFK Kennedy	LDJ	Linden NJ
		EWR Newark	FRG	Republic (Farmingdale)
			TEB	Teterboro NJ
			HPN	White Plains
TX	Dallas/ Fort Worth	DFW Regional	ADS	Addison
			DAL	Love
			FTW	Meacham

BIG AIRPORT SUMMARY

Most big airports are more trouble than they're worth to us little guys. Unless you have a major reason for being there, you're much better off at a satellite field. However, if you must, you must watch out for proper radio communications activity, jet blast, vehicles, and high prices.

When approaching the airport, use the ATIS, and Approach Control. They'll orient you better, and help you fit into the system. After you land, make sure you use the runway properly. Don't do anything the tower doesn't tell you to—you may have a 727 on your tail. Yes, causing a Concorde to pull up and go around is one of life's more embarrassing moments!

Table 7-2 gives traffic movements at the 25 busiest airports in the U.S.A. for 1980.

BIG AIRPORT QUIZ

Answers are on page 182.

1. You are VFR, approaching a major airline airport not located in a TCA. You should:

Table 7-2. Movements at Airports with FAA Control Towers Fiscal Year 1980.

Rank	Airport	State	Indent.	Movements (000)
1	Chicago, O'Hare*	IL	ORD	734.6
2	Long Beach, Daugherty	CA	LGB	645.2
3	Atlanta, Hartsfield*	GA	ATL	609.5
4	Santa Ana, Orange County	CA	SNA	569.8
5	Los Angeles, Van Nuys	CA	VNY	567.0
6	Los Angeles, International*	CA	LAX	534.4
7	Oakland, International	CA	OAK	487.6
8	Denver, Stapleton	CO	DEN	485.7
9	Dallas, Fort Worth, Regional*	TX	DFW	467.1
10	Miami, Opa Locka	FL	OPF	450.8
11	Miami, Tamiami	FL	TMB	422.9
12	San Jose, Municipal	CA	SJC	415.6
13	Seattle, Boeing	WA	BFI	408.2
14	Fort Worth, Meacham	TX	FTW	400.7
15	Denver, Arapahoe County	CO	APA	393.7
16	Phoenix, Sky Harbor	AZ	PHX	390.4
17	Honolulu, International	HI	HNL	385.5
18	Miami, International*	FL	MIA	376.8
19	San Francisco, International*	CA	SFO	371.2
20	Torrance, Municipal	CA	TOA	370.4
21	Las Vegas, McCarran	NV	LAS	364.4
22	Washington, National*	VA	DCA	354.7
23	Pittsburgh, Greater-	PA	PIT	353.1
24	Houston, Hobby	TX	HOU	350.4
25	Boston, Logan*	MA	BOS	340.9

*Major airline airport   Source: FAA

a. Contact the Tower within 15 miles for landing instructions.
b. Contact Approach Control, then listen to the ATIS.
c. Listen to the ATIS and contact Approach Control.
d. Contact the Tower within 5 miles for landing instructions.

2. Landing fees:
a. Are set by the FAA.
b. Vary by aircraft weight, airport, time of day, and type of operation.
c. Are not payable by non-revenue flights.
d. None of the above.

3. You want to go into a major airline field to drop off a passenger. You would expect:
a. To park at the regular airline gates.
b. To go through a security check before entering the terminal.
c. To have to go to the general-aviation facility, and get a ride to the terminal.
d. Not to be able to do this.

4. You are taxiing in your light aircraft behind a Lockheed 1011. How far behind it should you stay?
a. 2,100 feet
b. 850 feet
c. 350 feet
d. 150 feet

5. A flashing white light signal while waiting to take off means:
a. Return to the starting point on the airport.
b. Cleared for takeoff.
c. Taxi clear of the runway in use.
d. Taxi into position and hold.

6. A steady green signal in the air means:
a. Continue circling.
b. Cleared to land.

c. Cleared for touch and go.
d. Cleared for the option.

7. A flashing green signal in the air means:
a. Cleared for touch and go.
b. Return for landing.
c. Give way to other aircraft and continue circling.
d. Airport unsafe—do not land.

8. An alternating red and green signal on the ground means:
a. Exercise extreme caution.
b. Return to the starting point on the airport.
c. Taxi clear of the runway in use.
d. Taxi into position and hold.

9. A flashing white signal in the air means:
a. Exercise extreme caution.
b. Cleared for touch and go.
c. Return for landing.
d. None of the above.

10. Use of the ATIS is:
a. Mandatory.
b. Optional.

Chapter 8

Flying in Terminal Control Areas

The purpose of a Terminal Control Area (TCA) is to provide a high measure of control over—and thus, presumably, improved safety for—all traffic operating within the confines of certain high traffic-density areas. TCAs exist at major airline hubs throughout the U.S.A. Special pilot and equipment requirements exist for flight within TCAs.

Regardless of weather conditions, ATC authorization is required prior to operating within a TCA. You should not request such authorization unless you can meet the requirements mentioned above.

FLIGHT PROCEDURES: VFR FLIGHTS

If you are flying inbound VFR and want to enter the TCA, you should contact ATC on the appropriate frequency and in relation to geographical fixes shown on local charts (see appropriate radio facility or visual terminal area charts for details), or relative to radio aids in the area. Try to establish communications with ATC at least five minutes before you expect to enter the TCA, bearing in mind that you may *not* enter the TCA airspace until you have received a clearance to do so.

In establishing communications, what you have to do is tell the controller where you are (position and altitude), what type of aircraft you are flying, and state your request.

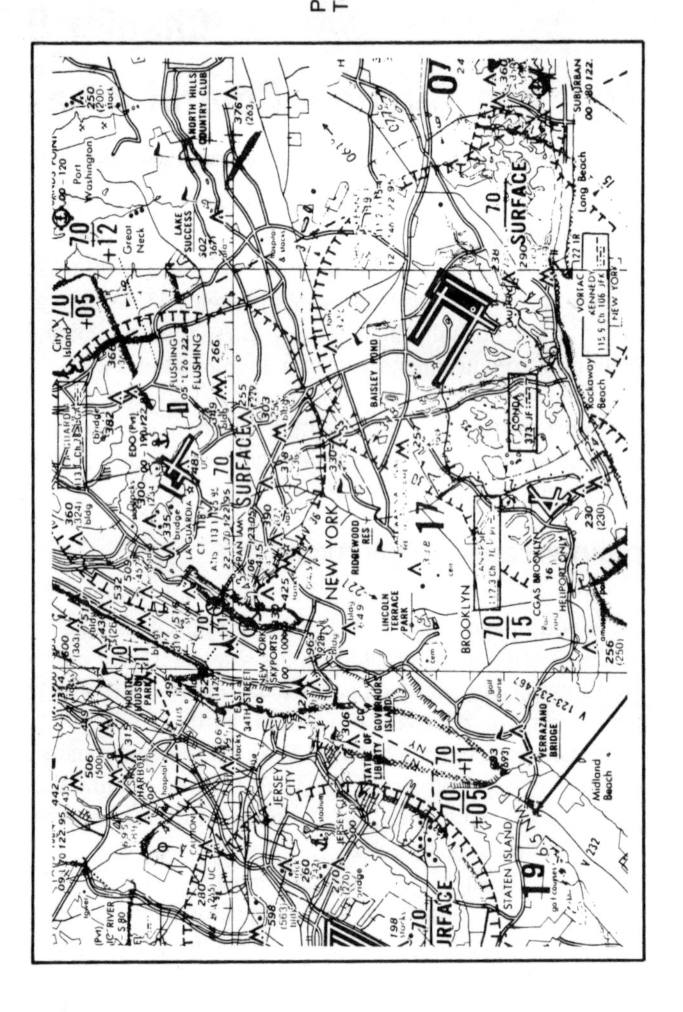

Part of the New York VFR Terminal Area chart.

For example, if you were flying from Trenton NJ to Newark, which is right in the New York TCA, you'd look at the TCA chart and see that you should contact New York Approach on 125.5 when approaching from the southwest. However, you should first get the ATIS report from Newark, since you should have received this before contacting Approach. So you look on the Newark approach chart for the ATIS frequency and find that it is broadcast on 115.7 mHz, which is Sparta VOR. Here's how the ATIS might sound:

"Newark arrival information Romeo. One five zero zero Greenwich weather, measured ceiling seven thousand overcast, visibility seven miles. Temperature six eight. Wind one two zero at ten. Altimeter two niner niner two. ILS runway four right approach in use. Landing runway four right. VFR arrivals may be assigned runway one one. Advise you have Romeo."

Having noted this information, you contact New York Approach. Here's how your conversation might go:

You: "New York Approach, this is Comanche N8251P."

ATC: "Comanche 8251P, go ahead."

You: "51P is 10 DME from Yardley on the 050° radial at 2,000 feet, squawking 1200, landing Newark, with Romeo [the latest ATIS report]."

First the controller will want to identify you on radar, so he'll give you your own discrete transponder code and ask you to push the ident button, which causes your radar target to bloom on his scope. He'll say:

ATC: "51P squawk 5205 and ident."

You: "51P, roger, 5205, ident."

You set your transponder to 5205, and press the ident button.

ATC: "51P radar contact. Cleared into the TCA on your present heading to maintain 2,500 feet, radar vectors to Newark runway 11, turn right to 050°."

You: "Roger 51P, cleared in to the TCA, leaving 2,000 for 2,500, turning to 050°."

You proceed inbound at 2,500 feet on the 050° heading given to you. As you get closer in, the controller will switch you to a local TCA controller in the Newark tower. (The controller you are presently talking to is in the New York Common IFR Room, located in Long Island City, New York, where he works alongside the approach controllers for Kennedy, La Guardia, and Westchester airports.)

ATC: "51P, descend to 1,500 feet. Contact Newark TCA on 127.85 now."

You: "51P, roger, leaving 2,500 for 1,500, 127.85."

You follow the clearance and contact Newark:

You: "Newark TCA, 8251P, out of 2,000 for 1,500."

ATC: "51P, radar contact, you'll be number two for one one, following a Navajo, he's two miles out on right base. Enter a right base and descend to 1,000 feet."

You: "51P to 1,000, to right base for one one."

You continue inbound, descending as cleared, and enter the right base leg for runway 11. You search for the traffic you were issued, and prepare the airplane for landing, carrying out your pre-landing checks.

ATC: "51P contact Newark tower 118.3"

You: "51P, 118.3"

You change frequency, and call the tower.

You: "Newark tower, 8251P, right base for one one."

ATC: "51P, cleared to land one one."

And that's all there is to it!

"Remain Clear of TCA"

All too often, at busy periods when you call for a TCA clearance, you will be told to stay clear of the TCA. All TCA operations are on a first-come, first-served basis, and it doesn't take much for the controller to become overloaded. The only thing ATC can do in such circumstances is turn off the faucet of incoming flights until things calm down a bit. That's great for them, but what about you? Airplanes can't stop in midair, and there are no VFR holding procedures established for this type of eventuality. You are on your own. Let's look at a typical situation, and how you can handle it to get what you want. Let's say you want to repeat the above exercise—land at Newark. Here's how the soundtrack might run. The controller is yapping away to all kinds of traffic. You wait for a quite moment, and call:

You: "New York Approach, this is Comanche N8251P."

No answer. That's the first sign of trouble. The controller continues talking to other traffic, and again you find a spot of silence and call:

You: "New York Approach, this is Comanche N8251P."

Obviously, judging by the hectic airwaves, the TCA is busy. You may hear another aircraft you haven't heard before call in for the first time and get a reply—always a frustration when ATC didn't answer you on your two calls. And this frustration is multiplied when the controller tells the other aircraft that there'll be at least a ten-minute delay. Keep your cool! First of all, the ten-minute delay is not a rigid matter, it's a guess by the controller—it may be more and it could easily be less.

Now, further contributing to your sense of being unwanted and unloved, *another* stranger calls in, gets a reply, and is told the same delay story. *Get in that line!* Call again!

You: "New York Approach, this is Comanche N8251P."

Finally, ATC hears you and replies:

ATC: "Comanche calling, remain clear of the TCA, call me again in ten minutes. All aircraft, there will be at least a ten- to fifteen-minute delay entering the TCA due to traffic congestion."

At least now you are closer to being in the lineup. But note that the controller did not call you by number—he doesn't want to clog up his personal short-term memory with your identification. The rule here is patience, perseverance and caution.

Slow the airplane down— you're not going anywhere in a hurry, and there's no sense in wasting fuel. Make sure you don't penetrate the TCA. If you get to the edge of it, start a lazy 360° turn, or plan your descent so you can get further toward your destination below the TCA. But whatever you do, don't enter that TCA! And another thing— since the sky is busy, guess what? Maybe there's some traffic around you! Double up your scanning efforts. Keep your eyes peeled and moving, searching for other airplanes. If you have passengers, get them to look, too. Maybe there are some others wanting to enter the TCA in your area who haven't even been able to start making contact yet.

Monitor the frequency, and when things seem to have calmed down a bit, give it another try. You will eventually succeed. And when you establish communications and you're ready to state your request, be accurate, brief, and to the point.

Transiting the TCA

So far we have looked at entering the TCA to land at one of its airports. What about when you want to just fly through? You will find, in busy periods, it's harder to get service for a transitory flight than for one that's landing within the TCA. The reasoning seems to be that you can always go around the TCA if you want to pass through, whereas if you want to land within it, you don't have much of an alternative.

The trick here is to get what you want, which you can only do without disrupting everyone else. That means you've got to fit in with the flow. For example, flying through the New York TCA from the southwest to the northeast means you will be encountering first Newark arrivals and departures, then La Guardia arrivals, then Kennedy arrivals and departures, then Kennedy arrivals again. This means there are certain altitudes you just won't be able to get for a direct routing.

Let's say you are flying from Yardley direct to La Guardia direct to Hartford, which is a straight line. The ideal way to do this is to get above the TCA. ATC has no jurisdiction over you in VFR conditions above the top of the TCA. The New York TCA tops out (at present) at 7,000 feet. You are going northeastwards, so you need a VFR altitude of "odds plus 500." The first altitude above the TCA is 7,500 and the next is 9,500.

If you can get above the TCA without entering it, this is the best bet. However, from a traffic point of view, in so doing you'll also be entering the let-down airspace going into New York's three airports, so it's a good idea to get in radar contact anyway.

You: "New York Approach, this is Comanche N8251P."

ATC: "Comanche 8251P, go ahead."

You: "51P is 12 DME from Yardley on the 052° radial out of 3,500 feet for 7,500, squawking 1200, proceeding direct La Guardia Above TCA, destination Hartford. Request traffic advisories."

If he's not busy, the controller will identify you on radar, giving you a transponder code, and asking you to ident. He'll say:

ATC: "51P squawk 5211 and ident."

You: "51P, roger, 5211, ident."

ATC: "51P radar contact. Remain clear of the TCA. Proceed over the TCA on your present heading. Report reaching 7,500 feet."

You: "51P."

You proceed as instructed, and will receive traffic advisories as well as the controller can supply them. Bear in mind that he is under no obligation to give you traffic, and if he's too busy, he may not. So, as always, keep your eyes peeled!

If he is busy, he may tell you on your first call-up:

ATC: "51P, unable traffic advisories, we're too busy. Stay clear of the TCA altimeter two nine nine two."

If you're smart you'll reply something like this:

You: "Roger 51P, proceeding above TCA as advised at 7,500 feet. Can you at least give a transponder code?"

You may or may not get one. If you do, you'll probably end up getting traffic advisories, too. The advantage of getting a code is that you will then become tagged on his radar scope, with your identification, speed and altitude (assuming you have an encoder—if you don't, get one!). Otherwise, you'll just be another VFR target, one of the great unwashed. Believe me, you're much better off being part of their system than being some strange nuisance that they don't know about.

Let's suppose that the ceilings are too low to enable

you to overfly the TCA. You have to go through, or around. If they're really busy, you're probably better off going around and not messing with them. But still try to get VFR traffic advisories. If you are able to go through, it'll probably be at an altitude of at least 4,000 feet above ground so you don't encroach on arrivals and departures at the main airports.

Don't Talk Too Much

The key point in communicating with TCA controllers is to keep every communication brief and succinct. If you're overflying, he doesn't want to know your route all the way to your destination. Just give the route to the first VOR on the other side of the TCA. Pilots who give their life histories can bung up the frequency, and preclude information getting out to the right people.

Let me give you a perfect example from recent experience. I was in the New York TCA, going from northeast to southwest at 6,500 feet. I was working La Guardia approach. Some guy calls up as follows:

"McGuire Approach, Cherokee 77 xray."

"77 xray, La Guardia."

First problem: The idiot had called up on the wrong frequency. He had the usual poor radio equipment found in too many light aircraft, and La Guardia—which he reached—thought he was saying "La Guardia" when he was, in fact, saying "McGuire."

"77xray, we're just off ah old bridge and we're heading down to ah Neshaminy Mall to get some ah photographs. We'll be, oh, I guess, at about 1,500 feet, and we're ah climbing through, ah, 1,000 now. We'd like, ah traffic advisories, and we're ah negative transponder. We're just going to ah hold this heading, unless you ah want me to turn for ah radar identification. We're a ah Cherokee 180, and we'll be landing at Burlington ah County Airport after the ah photo session."

Second problem: La Guardia has never heard of Neshaminy Mall, which is about 15 miles northeast of Philadelphia. And the request was inane from the point of view of New York TCA traffic—you can't deal with New

York TCA without a transponder, for example. No wonder poor old La Guardia was a little confused.

Meanwhile, there I was at 6,500 feet—my clearance altitude—watching an Eastern A300B descending directly toward me from above. He looked like he was coming right at me. What do I do? I can't get a word in edgewise while this clown is screwing up the busiest TCA in the world with his "wrong number" call. I decided to get out of the Airbus' way, fast, which I did by climbing and turning right. I reasoned that in 500 feet, I would be out of the TCA. Finally the fool stopped talking, and I cut in: "La Guardia, that aircraft thinks he's talking to McGuire Approach, and we've climbed to 7,500 feet to avoid an airbus, 8251P." The controller quickly dispatched the errant Cherokee to have his fun elsewhere, and then got mad at *me*.

ATC: "51P, did you climb just now?"

Me: "51P affirmative, we're at 7,500 feet. We climbed to avoid an airbus that was coming right at us."

ATC: "51P, he knew about you and was holding 7,000 feet until he was by you."

Me: "Well, he sure looked like he was coming right at me, so I took evasive action. Now we'd like to start our descent, please."

Seriously folks, what would *you* have done? An airbus coming at you at a closing speed of about 350 knots looks like it means business. If you can't talk because of an idiot, what are you supposed to do? My solution: exercise the prerogative of the pilot-in-command and act now! Ask (and answer, you can bet) questions later!

Departing a TCA from Within

If you are at an airport that lies within a TCA and you want to depart VFR, you will have to get a clearance to enter the TCA (which you will do as soon as you lift off the runway) before takeoff. You do this by contacting *Clearance Delivery* (if there is one) or Ground Control (if there isn't). Here is a typical sequence, once again using Newark.

If the airport is busy, you may want to do all of the preliminary work before you start engines so as to conserve fuel. If it looks like there'll be a delay, you're better off talking from battery power rather than running your engine needlessly.

First get your ATIS. Get the frequency from the airport chart and note the important points—departure runway, wind, altimeter, and so on. Then call Clearance Delivery:

You: "Newark Clearance, Comanche 8251P."

ATC: "51P, go ahead."

You: "8251P, we'd like a clearance to depart the TCA southwest, destination Washington National, requesting a climb to 6,500 feet."

ATC: "51P, clearance on request."

"Clearance on request" means that they'll get back to you. Sometimes, on initial call up, you might get this:

You: "Newark Clearance, Comanche 8251P."

ATC: "51P, clearance on request."

The controller is not being clairvoyant. He merely thinks you have an IFR flight plan on file, and it hasn't got to him yet. Set him straight:

You: "51P, we have no flight plan on file. We'd like a clearance to depart the TCA southwest, etc. etc."

Anyway, while the clearance is on request, you sit there and wait. In a few moments (usually), he'll get back to you with:

ATC: "51P, clearance."

Have your pencil ready and copy the key information as he gives it:

You: "51P, go ahead."

ATC: "51P is cleared out of the TCA to maintain 1,500 feet. Depart runway one one. After departure turn right heading 180°, radar vectors. Squawk 3926. Departure control frequency will be 119.2."

You: "Roger, 51P is cleared out of the TCA, maintain 1,500, off runway one one, turn right 180°, vectors, 3926, 119.2°."

ATC: "51P, clearance correct."

In my airplane I have installed an old thumbwheel switch I bought at a radio surplus store for $10 as a heading and altitude annunciator, so I don't have to do any writing when I get a TCA clearance. In the above situation, I set the heading bug on my HSI (horizontal situation indicator—you could use your DG, if it has a heading bug) to 110° for the runway, the heading annunicator thumbwheel to 180° for my first turn, the altitude annunciator to 1,500, the transponder to 3926, and number two COMM to 119.2. Then I simply read back the information from these settings.

Now you have your clearance, and assuming no delay (which would have been announced on the ATIS, or when you got your clearance), you shut your radios down (to avoid damage when starting) and start engines. Then contact Ground Control:

You: "Newark Ground, Comanche 8251P."

ATC: "51P, go ahead."

You: "8251P, at Butler with Oscar, VFR to Washington National, taxi clearance for runway 11."

You'll get conventional taxi instructions, and then switch to tower for takeoff clearance, as you would at any controlled field. After takeoff (make sure you turn your transponder on as you start your roll), you contact Departure Control when instructed by the tower.

You: "Newark Departure, 8251P with you, going through 700 feet."

ATC: "51P, radar contact, turn right to 200°, climb to 4,500 feet. Report reaching."

Note that as soon as you are in radar contact, you'll probably get a whole new set of instructions—different headings, different altitudes, etc. This is because the clearance was really just a script. Now you're on stage in the real world, and you'll do what seems best given the present traffic situation. This can mean getting an altitude restriction, or the removal of an altitude restriction, and an expedited turn to your direct routing.

After you've left the TCA, the controller will release you, as follows:

ATC: "51P, you're leaving the TCA and my airspace. No further traffic observed on your route. Squawk the appropriate VFR code, frequency change approved."

By "the appropriate VFR code," he means 1200. If you want to continue receiving radar advisories (which I recommend), ask for a frequency:

You: "Roger, 51P, thank you. Can you suggest a frequency for further advisories?"

ATC: "Try Philadelphia Approach on 123.8."

In fact, on this route, you'll probably have to fly through the Philly TCA, so now you'll go through the process of getting a new TCA clearance for a transitory flight, or else climb quickly to get above their TCA before you get to its boundary.

And so it goes.

Departing a TCA from an Airport outside the TCA

If you are taking off from an airport that is adjacent to the TCA, or underlying it (e.g., Teterboro, or Linden, New Jersey), and you want to get a clearance through the TCA, you're best off to get it in the air. If you try to get a TCA clearance on the ground, you could get a long delay. After you are airborne, contact the appropriate frequency and ask for a TCA clearance as you would for a transitory flight. As always, remain clear of the TCA until you have a clearance through.

FLIGHT PROCEDURES: IFR FLIGHTS

Flying IFR in a TCA is just like flying IFR in any controlled airspace, with one exception. If you are in a TCA, and you want to *cancel IFR,* you must obtain a VFR clearance before cancelling. In some cases, it may be better not to cancel, since you are already in the system when you are IFR, and to convert to VFR can cause unnecessary complications.

RADIO FAILURE IN A TCA

Radio failure is an event that can creep up on you without you realizing it for some time after its happened. You don't usually get a big bang and flash, as in the movies. It's just that things get quiet, and you gradually realize that no one has said anything for a while. If you hear silence for

The high-efficiency Mooney 201. (courtesy Mooney)

more than one minute during an otherwise fairly busy period, make a call. If they reply, just indicate that you hadn't heard anything for a while, and you are "just checking." If you start calling, and don't get any replies, maybe you have a problem. Here's what to do if the radio seems dead:

First, check the squelch adjustment and volume control on your COMM. Momentarily press the mike button and see if you get an indication of transmission. If those don't produce results, try another frequency, such as an ATIS, to see if you can hear anything. If that doesn't work, try another radio. Try another mike. Are your NAV radios working? Is your transponder light blinking, indicating that it's being interrogated? If so, you can communicate by using the ident. button, since every time you push it, ATC can see that on their radar.

If you have lost electrical power, everything else that is electrically operated will also be dead—navigation lights, gear/flap operating mechanisms, turn indicator, autopilot, and so on. Turn off unnecessary electrical devices and try to restore electrical power: Check your generator/alternator. Look at the ammeter or load meter.

Look at the voltmeter, if you have one. Check all the circuit breakers.

If you can't correct the situation, and you are IFR, follow the procedures in FAR 91.127. If you're VFR in a TCA, set your transponder to 7700 (emergency) for one minute (this causes warnings to flash on the radar screens). Then set it to 7600 (radio failure).

Keep listening; ATC will call you, and if you receive, will have you squawk ident to acknowledge instructions, for example:

ATC: "8251P, if you read squawk ident."

You press the ident button.

ATC: "51P, roger, descend to 2,500 feet and leave the TCA. Squawk ident to acknowledge."

You do so, and get out of their hair.

If you can't receive, proceed out of the TCA at your last assigned altitude, maintaining VFR. Go to the nearest suitable airport and land. Then call the ATC agency you were working and advise them what happened. If you have to land at a controlled airport without radio, remain outside or above the airport traffic area until you figure out the direction or flow of traffic then join the airport traffic pattern and maintain visual contact with the tower to receive light signals. If your receiver is working, listen on the tower frequency for instructions, and respond by rocking your wings or flashing your landing or navigation lights. If your transmitter is working, but your receiver isn't, follow the above pattern-joining process, and transmit blind, looking for light signals in response.

GETTING WEATHER INFORMATION

As well as the local ATIS, broadcasts over NAV stations and calls to the FSS, in many TCAs you can get weather information by calling Flight Watch on the standard frequency of 122.0.

SUMMARY

Knowing how to fly in TCAs and feeling comfortable doing so is the major portion of your battle of flying in congested airspace. The secret is maintaining a level of awareness while you are in the TCA. Listen to all the other aircraft; gauge what's going on. Move with the flow; don't try to start your own system. Be brief and professional in your communications.

If you live near a TCA, and you don't feel comfortable now, get a VHF receiver and spend some time listening on the TCA frequency. You'll be amazed at how soon you'll get it together.

TCA QUIZ

Answers are on page 182.

1. ATC authorization is required before operating in a TCA:

a. At all times.
b. In IFR conditions only.
c. In VFR conditions only.
d. In Group I TCAs only.

2. A transponder is required for operation in:

a. Group I and Group II TCAs only.
b. Group I TCAs only.
c. Group II TCAs only.
d. Group III TCAs only.

3. An encoding altimeter is required for operation in:

a. Any TCA.
b. Group I TCAs only.
c. Group II TCAs only.
d. Group III TCAs only.

4. DME is required for operation in:

a. Group I TCAs only.
b. Group II TCAs only.
c. Group III TCAs only.
d. None of the above.

5. ATC authorization to overfly a TCA above its top limit is:

a. Always required.
b. Not required.
c. Required of Class I TCAs only.
d. Required of Class II TCAs only.

Chapter 9

Don't Let This Happen to You!

About 0901:47 PST, September 25, 1978, Pacific Southwest Airlines Flight 182, a Boeing 727, and a Gibbs Flite Center 172 collided in midair about three nautical miles northeast of Lindbergh Field, San Diego, California.

The Cessna was under the control of San Diego Approach Control and was climbing on a northeast heading. PSA Flight 182 was making a visual approach to runway 27 at Lindbergh Field and had been advised of the location of the Cessna by the Approach Controller.

The flight crew told the Approach Controller that they had the traffic in sight and were instructed to maintain visual separation from the Cessna and to contact Lindbergh Tower. PSA 182 contacted the tower on its downwind leg and was again advised of the Cessna's position. The flight crew did not have the Cessna in sight; they thought they had passed it and continued the approach. The aircraft collided near 2,600 feet MSL and fell to the ground in a residential area. Both occupants of the Cessna were killed; 135 persons on board the Boeing 727 were killed; seven persons on the ground were killed; nine persons on the ground were injured. Twenty-two dwellings were damaged or destroyed. The weather was clear, and the visibility was ten miles.

The National Transportation Safety Board determined that the probable cause of the accident was the failure of the flight crew of PSA 182 to comply with the provisions of a maintain-visual-separation clearance, including the requirement to inform the controller when they no longer had the other aircraft in sight.

Contributing to the accident were the air traffic control procedures in effect, which authorized the controllers to use visual separation procedure to separate two aircraft on potentially conflicting tracks when the capability was available to provide either lateral or vertical radar separation to either aircraft.

BACKGROUND OF THE FLIGHT

About 0816 PST (all times herein are Pacific Standard based on the 24-hour clock) air traffic control procedures were in effect, which authorized the controllers to use visual separation procedures to separate two aircraft on potentially conflicting tracks.

A flight instructor occupied the right seat of Cessna N711G, and another certificated pilot, who was receiving instrument training, occupied the left seat.

The Cessna proceeded to Lindbergh Field, where two practice ILS approaches to Runway 9 were flown. Although the reported wind was calm, Runway 27 was the active runway at Lindbergh. About 0857, N7711G ended a second approach and began a climbout to the northeast; at 0859:01, the Lindbergh tower controller cleared the Cessna pilot to maintain VFR conditions and to contact San Diego Control.

At 0859:50, the Cessna pilot contacted San Diego Approach Control and stated that he was at 1,500 feet, (all altitudes herein are above mean sea level unless otherwise specified) and "northeast-bound." The Approach Controller told him that he was in radar contact and instructed him to maintain VFR conditions at or below 3,500 feet and to fly a heading of 070 degrees. The Cessna pilot acknowledged and repeated the controller's instruction.

Pacific Southwest Airlines Flight 182 was a regularly scheduled passenger flight between Sacramento and San Diego, California, with an intermediate stop in Los

Angeles. The flight departed Los Angeles at 0834 on an IFR flight plan with 128 passengers and a crew of seven on board. The first officer was flying the aircraft. Company personnel familiar with the pilots' voices identified the captain as the person conducting almost all air-to-ground communications. The cockpit voice recorder (CVR) established the fact that a deadheading company pilot occupied the forward observer seat in the cockpit.

At 0853:19, Flight 182 reported to San Diego Approach Control at 11,000 feet and was cleared to descend to 7,000 feet. At 0857, Flight 182 reported that it was leaving 9,500 feet for 7,000 feet and that the airport was in sight. The Approach Controller cleared the flight for a visual approach (an approach wherein an aircraft on an IFR flight plan operating in VFR conditions under control of an ATC facility and having an ATC authorization may proceed to the airport of designation in VFR conditions) to Runway 27; Flight 182 acknowledged and repeated the approach clearance.

At 0859:28, the Approach Controller advised Flight 182 that there was "Traffic 12 o'clock, one mile, northbound." Five seconds later the flight answered, "We're Looking."

At 0859:39, the Approach Controller advised Flight 182, "Additional traffic 12 o'clock, three miles, just north of the field, northeastbound, a Cessna 172 climbing VFR out of one thousand four hundred." According to the CVR, at 0859:50, the copilot responded, "Okay, we've got that other twelve."

At 0900:15, about 15 seconds after instructing the Cessna pilot to maintain VFR at or below 3,500 feet and to fly 070 degrees, the Approach Controller advised Flight 182 that "Traffic's at 12 o'clock, three miles, out of one thousand seven hundred."

At 0900:21, the first officer said "Got em", and one second later the captain informed the controller, "Traffic in sight."

At 0900:23, the Approach Controller cleared Flight 182 to "Maintain visual separation," and to contact Lindbergh tower. At 0900:28 Flight 182 answered "Okay," and three seconds later the Approach Controller advised the Cessna pilot that there was "Traffic at 6 o'clock, two miles, eastbound; a PSA jet inbound to Lindbergh, out of three thousand two hundred, has you in sight." The Cessna pilot acknowledged, "One one golf, roger."

At 0900:34, Flight 182 reported to Lindbergh tower that they were on the downwind leg for landing. The tower acknowledged the transmission and informed Flight 182 that there was "Traffic, 12 o'clock, one Mile, a Cessna."

At 0900:41, the first officer called for five degree flaps, and the captain asked, "Is that the one (we're) looking at?" The first officer answered, "Yeah, but I don't see him now." According to the CVR, at 0900:44, Flight 182 told the local controller, "Okay, we had it there a minute ago," and six seconds later, "I think he's passed off to our right." The local controller acknowledged the transmission. (According to the ATC transcript the 0900:50 transmission "Think he's passing off to our right" and the local controller testified that he heard, "He's passing off to our right.")

The CVR showed that Flight 182's flight crew continued to discuss the location of the traffic. At 0900:52, the captain said, "He was right over there a minute ago." The first officer answered, "Yeah."

At 901:21, the captain said, "Oh yeah, before we turned downwind, I saw him about one o'clock, probably behind us now."

At 0901:31, the first officer called, "Gear down."

At 0901:28, the conflict alert warning began in the San Diego Approach Control Facility, indicating to the controllers that the predicted flight paths of Flight 182 and the Cessna would enter the computer's prescribed warning parameters.

At 0901:47, the Approach Controller advised the Cessna pilot of "Traffic in your vicinity, a PSA jet has you in sight, he's descending for Lindbergh." The transmission was not acknowledged. The Approach Controller did not inform Lindbergh tower of the conflict alert involving Flight 182 and the Cessna, because he believed Flight 182's flight crew had the Cessna in sight.

The aircraft collided at 0901:47.

According to the witnesses, both aircraft were proceeding in an easterly direction before the collision. Flight 182 was descending and overtaking the Cessna, which was climbing in a wing-level attitude. Just before impact, Flight 182 banked to the right slightly, and the Cessna pitched nose-up and collided with the right wing of Flight 182.

The Cessna broke up immediately and exploded. Segments of fragmented wreckage fell from the right wing and empennage of Flight 182.

Flight 182 began a shallow right descending turn, leaving a trail of vaporlike substance from the right wing. A bright orange fire erupted in the vicinity of the right wing and increased in intensity as the aircraft descended. The aircraft remained in the right turn, and both the bank and pitch angles increased during the descent to about 50 degrees at impact.

Both aircraft were destroyed by the collision, inflight and postimpact fires and impact. There were no survivors. Seven persons on the ground were killed, and 22 dwellings were damaged or destroyed.

FINDINGS

Flight 182 was cleared for a visual approach to Runway 27 at Lindbergh Field.

The Cessna was operating in an area where ATC control was being exercised, and its pilot was required either to comply with the ATC instruction to maintain the 070-degree heading or to advice the controller if he was unable to do so.

The Cessna pilot failed to maintain the assigned heading contained in his ATC instruction.

The cockpit visibility study shows that if the eyes of the Boeing 727 pilot were located at the aircraft's design eye reference point, the Cessna's target would have been visible.

Two separate air traffic control facilities were controlling traffic in the same airspace.

The Approach Controller did not instruct Flight 182 to maintain 4,000 feet until clear of the Montgomery Field airport traffic area in accordance with established procedures contained in Miramar Order NKY.206G.

The issuance and acceptance of the maintain-visual-separation clearance made the flight crew of Flight 182 responsible for seeing and avoiding the Cessna.

The flight crew of Flight 182 lost sight of the Cessna and did not clearly inform controller personnel of that fact.

The tower local controller advised Flight 182 that a Cessna was at 12 o'clock, one mile. The flight crew comments to the local controller indicated to him that they had passed or were passing the Cessna.

The traffic advisories issued to Flight 182 by the Approach Controller at 0900:15 and by the local controller at 0900:38 did not meet all the requirements of paragraph 511 of Handbook 7110.65A.

The Approach Controller received a conflict alert on Flight 182 and the Cessna at 0901:28. The conflict warning alerts the controller to the possibility that, under certain conditions, less-than-required separation may result if action is not, or has not been, taken to resolve the conflict. The Approach Controller took no action upon receipt of the conflict alert, because he believed that Flight 182 had the Cessna in sight and the conflict was resolved.

The conflict alert procedure in effect at the time of the accident did not require that the controller warn the pilots of the aircraft involved in the conflict situation.

Both aircraft were receiving Stage II terminal radar services. Flight 182 was an IFR aircraft; the Cessna was a participating VFR aircraft. Proper Stage II services were afforded both aircraft.

Stage 2 terminal service does not require that either lateral or vertical traffic separation minima be applied between IFR and participating VFR aircraft; however, the capability existed to provide this type separation to Flight 182.

The Boeing 727 probably was not controllable after the collision.

PROBABLE CAUSE

The National Transportation Safety Board determined that the probable cause of the accident was the failure of the flight crew of Flight 182 to comply with the provisions of a maintain-visual-separation clearance, including the requirements to inform the controller when they no longer had the other aircraft in sight.

Contributing to the accident were the air traffic control procedures to use visual separation procedures to separate two aircraft on potentially conflicting tracks when the capability was available to provide either lateral or vertical radar separation to either aircraft.

SAFETY RECOMMENDATIONS

As a result of this accident, the National Transportation Safety Board has recommended that the Federal Aviation Administration:

"Implement a Terminal Radar Service Area (TRSA) at Lindbergh Airport, San Diego, California. (Class 1—Urgent Action) (A-78-77)" "Review procedures at all airports which are used regularly by air carrier and general aviation aircraft to determine which other areas require either a terminal control area or a terminal control radar service area and establish the appropriate one. (Class 2—Priority Action) (A-78-78)"

"Use visual separation in terminal control areas and terminal radar service areas only when a pilot requests it, except for sequencing on the final approach with radar monitoring. (Class 1—Urgent Action) (A-78-82)"

"Re-evaluate its policy with regard to the use of visual separation in other terminal areas. (Class 2—Priority Action) (A-78-83)"

YOU CAN'T RELY ON ATC

At 12:10 on May 18, 1978, N6423K, a Cessna 150, and N121GW, a Falcon Fan Jet, collided in midair about 3½ miles west of Memphis, Tennessee. At the same time of the collision both aircraft were operating under the control jurisdiction of Memphis Tower at an assigned altitude of 2,000 feet MSL and were in radio/radar contact with different facility controllers on separate radio frequencies.

Visual meteorological conditions (VMC) prevailed at the time.

Investigation disclosed that the Cessna 150 was a VFR arrival from the west and was receiving Stage III radar service; the Falcon was operating in a closed traffic pattern on an IFR flight plan and was conducting multiple ILS approaches to runway 17R. Further investigation revealed that ATC failed to effect the required separation minima applicable to known VFR and IFR traffic operating within the designated TRSA because controller personnel responsible for the control of the two aircraft did not coordinate the particular operation being conducted with each other. As a result of this lack of coordination, neither of the two controllers handling the Falcon had any knowledge that the 150 was inbound traffic, and the third controller, who was providing control service to the 150, had no knowledge of the Falcon's traffic pattern operation within his airspace at 2,000 feet. Therefore, no one recognized that a conflict existed until the two aircraft were seen on radar about one mile apart. At that point, insufficient time was available for corrective action.

The NTSB is concerned that a single coordination procedural error effectively negated the control capability of an ATC system which used modern automated radar equipment and procedural concepts. NTSB examined facility procedures, automated equipment, and TRSA requirements carefully to determine if additional safeguards are feasible, and how such measures would have prevented this accident.

They concluded that there are two problem areas worthy of corrective action. The first involves the local operating procedures used at Memphis for close traffic pattern IFR operations, and the second involves the current rules for aircraft operations in a TRSA and related transponder requirements.

The NTSB report made the following observations and recommendations:

Both aircraft were being controlled in accordance with prescribed procedures and standard practices at an assigned altitude of 2,000 feet. The airspace within a five-mile radius of the airport, from the surface to 2,000 feet, is designated for and utilized by the facility for local control operations. Thus, responsibility for the control of air traffic within that airspace is the responsibility of the local controllers (LC 1 and 2). To effect procedural control, the LC-1 controller is responsible for traffic operating in the east and west quadrants of a five-mile circle around the airport which is formed by bisecting lines NW/SE and NE/SW that pass through the center of the airport. The LC-2 controller is responsible for traffic operating in the north and south quadrants. Any traffic operating in a close traffic pattern at 2,000 feet or below will traverse the airspace of both the LC-1 and LC-2 controllers. Every circuit of the close traffic pattern for RWY 17R at Memphis required coordination between the LC-2 and LC-1 controllers to acquire knowledge of mutual traffic and potential conflicts. Also, these controllers are obligated to separate traffic in accordance with applicable criteria for TRSA traffic.

The Safety Board believes that close traffic pattern operations at Memphis International Airport should be discontinued within the designated airspace for local control operations. The additional workload imposed on local controllers by the requirement to coordinate and effect Stage III/IFR separation minima between these aircraft comprises their ability to perform their primary duties. Since the physical layout of the Memphis Airport and control procedures utilized by the facility are somewhat unique, the Safety Board believes that ideally any closed traffic pattern operation wherein the aircraft will be executing multiple ILS approaches should be conducted at an assigned altitude of 2500 feet or above. Appropriate radar control personnel in the TRACON are better suited to provide radar separation service than the local controller. Accordingly, control responsibility should be transferred to Memphis TRACON.

The Safety Board is extremely concerned by existing requirements for an aircraft transponder for flight operations in certain designated controlled airspace. A transponder with altitude encoder is required for flight operations conducted above 12,500 feet MSL and within designated Group 1 TCAs. Group 2 TCAs also require a transponder, but no altitude encoder is needed. Such equipment is not required for flight within a designated TRSA, nor is there any requirement that a pilot establish radio contact with ATC when traversing a TRSA. Based on its investigation of this accident, the Safety Board concluded that the transponder requirements for flight operations within a TRSA and TCA II should be revised. In view of the ever-increasing availability of ATC automated equipment and the future development of Beacon Collision Avoidance System, Discreet Address Beacon System and Automated Traffic Advisory and Resolution Service (ATARS), it indicated that it would be untenable to fail to reevaluate the operational benefits and safety enhancement that the altitude encoding Mode C transponder provide in TRSA and TCA II operations.

It was evident to the Board that if an operating transponder had been installed aboard Cessna 6423K, identification of that aircraft with altitude data would most likely have been detected by controller personnel, and the accident would not have occurred. At locations where the conflict alert system is operational, a Mode C transponder provides another safeguard that serves to prevent the type of accident that occurred at Memphis.

With respect to those civil airports that have a designated TRSA with Stage III service provided, the Safety Board recognized that traffic operations differ greatly between such airports as Phoenix, Arizona, and Roanoke, Virginia. Because some of the larger airports, such as Phoenix, now generate high volume traffic which closely approximates the criterion used for the establishment of a TCA II, it believed that TRSA locations should be classified into two groups based upon traffic count and carrier operations. Like the TCAs, they could be classified as TRSA I and TRSA II locations. It is believed that TRSA I locations with the higher volume traffic and ATC automation should require a Mode C transponder to conduct flight operations within the TRSA, and VFR aircraft operating enroutine through the TRSA to establish radio communication with ATC before entering the TRSA.

Because of the large number of transponder-equipped aircraft that operate from the airports affected by the change in existing transponder requirements recommended, it believed such action feasible, timely, and justified in the interest of safer flight operations.

Accordingly, the NTSB recommended that the FAA:

1. Evaluate the close traffic pattern operations conducted at Memphis International Airport and consider establishment of a procedure whereby high-performance or turbine jet aircraft conducting multiple approaches for training purposes be assigned an altitude of 2,500 feet or above, which would place responsibility for control of the aircraft with TRACON personnel.

2. Evaluate operational data for each TRSA location and establish two categories of TRSAs. Those locations handling the largest volume of traffic with automated ATC equipment available should be designated TRSA I locations. The remaining areas would be designated TRSA II locations.

3. Require Mode C transponder equipment for operations within a TRSA I and Group II TCA and require that a pilot of a VFR flight traversing a TRSA I establish radio contact with the appropriate ATC facility before entering the designated airspace.

Appendix A

Terminal Control Area Airport Selection Guide

The following listings show all TCAs (except Honolulu) and give the major airline airports, the IFR and VFR reliever airports, and nearby military airfields for each location, organized first by the state in which the main TCA airport is located, then by city. Each non-military airport is preceded by its three/four letter/digit FAA identification code.

The Flight Service Station (FSS) that covers the TCA is indicated in the TCA header line, and the airport at which the FSS is located is indicated by the abbreviation FSS for that airport's listing.

The availability of a control tower or Unicom is shown for each airport. Not all towers operate 24 hours a day. (Check current sectionals or IFR charts for correct frequencies and operating hours.)

Military facility abbreviations are:

AAF —Army Air Field
AFB —Air Force Base
ANGB—Air National Guard Base
CGAS —Coast Guard Air Station
MCAS—Marine Corps Air Station
NAS —Naval Air Station
OLF —Outlying Field (Navy)

CALIFORNIA

TCA: Los Angeles (Group I) — FSS: Los Angeles

Main Airline Airports		
BUR Burbank	Tower	
LAX Los Angeles International	Tower	FSS
Other IFR Airports		
EMT El Monte	Tower	
FUL Fullerton	Tower	
HHR Hawthorne	Tower	
LGB Long Beach	Tower	
SMO Santa Monica	Tower	
TOA Torrance	Tower	
VNY Van Nuys	Tower	
VFR Airports		
CPM Compton	Unicom	
SFR San Fernando	Unicom	
WHP Whiteman	Unicom	
Military Airfields		
Los Alamitos ANGB	Tower	

TCA: San Diego (Group II) — FSS: San Diego

Main Airline Airport		
SAN San Diego - Intl-Lindbergh	Tower	FSS
Other IFR Airports		
CRQ Carlsbad - McClellan-Palomar	Unicom	
SDM San Diego - Brown Field Mun.	Tower	
MYF San Diego - Montgomery	Tower	
SEE San Diego (El Cajon) - Gillespie	Unicom	
L39 Ramona	Unicom	
VFR Airports		
L54 Agua Caliente Springs	Unicom	
Military Airfields		
Imperial Beach OLF	Tower	
Miramar NAS	Tower	
North Island NAS	Tower	

TCA: San Francisco (Group I) **FSS: Oakland**

Main Airline Airports			
SFO	San Francisco - International	Tower	
OAK	Oakland - Metro International	Tower	FSS
Other IFR Airports			
CCR	Concord - Buchanan	Unicom	
HWD	Hayward - Air Terminal	Unicom	
LVK	Livermore - Municipal	Unicom	
VFR Airports			
O12	Antioch	Unicom	
Q59	Fremont	Unicom	
HAF	Half Moon Bay	Unicom	
O56	Novato - Gnoss	Unicom	
PAO	Palo Alto - Santa Clara County	Unicom	
SQL	San Carlos	Unicom	
Military Airfields			
	Alameda NAS	Tower	
	Moffett NAS	Tower	

COLORADO

TCA: Denver (Group II) **FSS: Denver**

Main Airline Airport			
DEN	Denver - Stapleton Intl	Tower	FSS
Other IFR Airports			
APA	Denver - Arapahoe County	Unicom	
BJC	Denver - Jeffco	Unicom	
VFR Airports			
01V	Aurora - Columbine	Unicom	
1V5	Boulder - Municipal	Unicom	
Military Airfields			
	Buckley ANGB	Tower	

DISTRICT OF COLUMBIA
(VIRGINIA/MARYLAND) **FSS:**
TCA: Washington (Group I) **Washington***

Main Airline Airports			
DCA	Washington - National	VA	Tower
IAD	Washington - Dulles Intl	VA	Tower
BWI	Baltimore -/Washington Intl	MD	Tower

*FSS is not located on an airport. It is in Leesburg, VA.

Other IFR Airports

MTN	Baltimore - Glenn L. Martin	MD	Unicom
W10	Manassas - Municipal/Davis	VA	Unicom

VFR Airports

6W9	Arcola - Glascock	VA	Unicom
1W2	Baltimore	MD	Unicom
W48	Baltimore - Essex Skypark	MD	Unicom
W32	Clinton - Hyde	MD	Unicom
CGS	College Park	MD	Unicom
GAI	Gaithersburg - Montgomery Cy	MD	Unicom
W18	Laurel - Suburban	MD	Unicom
W09	Leesburg - Municipal/Godfrey	VA	Unicom
WOO	Mitchellville - Freeway	MD	Unicom

Military Airfields

	Andrews AFB	MD	Tower
	Fort Belvoir - Davison AAF	VA	Tower
	Fort Meade - Tipton AAF	MD	Tower

FLORIDA

TCA: Miami (Group I) **FSS: Miami**

Main Airline Airports

FLL	Fort Lauderdale - Hollywood Int'	Tower	
MIA	Miami - International	Tower	

Other IFR Airports

TNT	Miami - Dade Collier Training	Tower	
FXE	Fort Lauderdale - Executive	Tower	
OPA	Opa Locka	Tower	
TMB	Tamiami	Unicom	FSS

VFR Airports

BCT	Boca Raton - Public	Unicom	
HWO	Hollywood - North Perry	Tower	
X51	Homestead - General Aviation	Unicom	
X46	Opa Lock West		

Military Airfields

	Homestead AFB	Tower	

GEORGIA

TCA: Atlanta (Group I) **FSS: Atlanta**

Main Airline Airport

ATL	Atlanta - Hartsfield Intl	Tower	

Other IFR Airports

PDK	Atlanta - Dekalb Peachtree	Tower	
FTY	Atlanta - Fulton Cty/C. Brown	Tower	FSS
8A4	Marietta - McCollum/Cobb Cty	Unicom	

VFR Airports

8A9	Fairburn - South Fulton	Unicom	
9A7	Jonesboro - South Expressway	Tower	

Military Airfields

	Dobbins AFB/Atlanta NAS	Tower	

ILLINOIS

TCA: Chicago (Group I) **FSS: Chicago**

Main Airline Airports

MDW	Chicago - Midway	Tower	
ORD	Chicago - O'Hare Intl.	Tower	

Other IFR Airports

ARR	Aurora - Municipal	Tower	
3HA	Chicago/Lansing - Municipal	Unicom	
DPA	Chicago/W. Chicago - DuPage	Tower	FSS
PWK	Chicago/Wheeling - Pal-Waukee	Tower	
C06	Elgin	Unicom	
JOT	Joliet - Park District	Unicom	

VFR Airports

3HW	Chicago/Blue Island - Howell	Unicom	
CGX	Chicago - Meigs	Tower	
06C	Chicago/Schaumburg	Unicom	

Military Airfields

	Glenview NAS	Tower	

LOUISIANA

TCA: New Orleans (Group II) **FSS:**

Main Airline Airport

MSY	New Orleans - Intl/Moisant	Tower	

Other IFR Airports

NEW	New Orleans - Lakefront	Tower	FSS
6R0	Slidell	Unicom	

VFR Airports

LA08	Covington - Privette	Unicom	

Military Airfields

	New Orleans NAS	Tower	

MASSACHUSETTS

TCA: Boston (Group I) **FSS: Boston**

Main Airline Airport

BOS	Boston - Logan Intl	Tower	FSS

Other IFR Airports

BED	Bedford - Hanscom	Tower	
BVY	Beverly - Municipal	Tower	
8B6	Haverhill	Unicom	
2B2	Newburyport - Plum Island	Unicom	
OWD	Norwood - Municipal	Tower	
6B6	Stow - Minute Man	Unicom	
BO9	Tewksbury - Tew Mac	Unicom	

VFR Airports

9B1	Marlboro	Unicom	
3B2	Marshfield	Unicom	

Military Airfields

	Weymouth NAS	Tower	

MICHIGAN

TCA: Detroit (Group II) **FSS: Detroit**

Main Airline Airport

DTW	Detroit - Metro Wayne County	Tower	

Other IFR Airports

DET	Detroit - City	Tower	FSS
2G5	Detroit - Grosse Ile Municipal	Unicom	
1D2	Plymouth - Mettetal	Unicom	
PTK	Pontiac - Oakland	Tower	
D98	Romeo	Unicom	
7D2	Troy - Oakland	Unicom	
YQG	Windsor (Ontario Canada)	Tower	
YIP	Ypsilanti - Willow Run	Tower	

VFR Airports

D13	Fraser - McKinley	Unicom	
3BB	Troy - Big Beaver	Unicom	

Military Airfields

	Selfridge AFB	Tower	

MINNESOTA

TCA: Minneapolis (Group II) **FS: Minneapolis**

Main Airline Airport

MSP	Minneapolis -/St Paul Intl	Tower	FSS

Other IFR Airports

Y25	Anoka - Gateway N. Industrial	Unicom	
8Y2	Buffalo - Municipal	Unicom	
ANE	Minneapolis - Anoka Cty. Blaine	Tower	
MIC	Minneapolis - Crystal	Tower	
FCM	Minneapolis - Flying Cloud	Tower	
STP	St. Paul - Downtown/Holman	Tower	
21D	St. Paul - Lake Elmo	Unicom	
D97	South St. Paul - Mun./Fleming	Unicom	

VFR Airports

Y12	Lakeville - Airlake Industrial	Unicom	

Military Airfields

None

MISSOURI

Kansas City (Group II) **FSS: Kansas City**

Main Airline Airport

MCI	Kansas City International	MO	Tower	

Other IFR Airports

3EX	Excelsior Springs - Memorial	MO	Unicom	
3GV	Grain Valley - E. Kansas City	MO	Unicom	
MKC	Kansas City - Downtown	MO	Tower	FSS
KCK	Kansas City - Fairfax	KS	Tower	
GVW	Kansas City - Richards-Gebaur	MO	Tower	
FLV	Leavenworth - Sherman AAF	KS	Tower	
IXD	Olathe - Johnson Cty. Ind.	KS	Tower	
OJC	Olathe - Johnson Cty. Exec.	KS	Tower	

VFR Airports

MO06	Kansas City - Heart	MO	Unicom	
51K	Olathe - Cedar	KS	Unicom	

Military Airfields

See Leavenworth (above)

TCA: St. Louis (Group II) **FSS: St. Louis**

Main Airline Airport

STL	St. Louis - Lambert Intl.	MO	Tower	

Other IFR Airports

ALN	Alton - Civic Memorial	IL	Unicom	
CPS	East St. Louis/BiState Parks	IL	Tower	
H22	Festus - Memorial	MO	Unicom	
3SQ	St. Charles	MO	Unicom	
3SZ	St. Charles - County	MO	Unicom	
02K	St. Louis - Arrowhead	MO	Unicom	
1H0	St. Louis - Creve Coeur	MO	Unicom	
SUS	St. Louis - Spirit of	MO	Tower	FSS
3WE	St. Louis - Weiss	MO	Unicom	

VFR Airports

None

Military Airfields

Scott AFB	Tower	

NEVADA

TCA: Las Vegas (Group II) **FSS: Las Vegas**

Main Airline Airport

LAS	Las Vegas - McCarran Intl.	Tower	FSS

Other IFR Airports

None

VFR Airports

BLD	Boulder City Municipal	Unicom	
L15	Las Vegas - Henderson/Sky Harbor	Unicom	
VGT	Las Vegas - North/Air Terminal	Tower	

Military Airfields

Nellis AFB	Tower	

NEW YORK

TCA: New York (Group I) **FSS: New York Teterboro**

Main Airline Airports

JFK	New York - Kennedy Intl	NY	Tower	
LGA	New York - La Guardia	NY	Tower	
EWR	Newark - International	NJ	Tower	

Other IFR Airports

N64	Basking Ridge - Somerset Hills	NJ	Unicom	
CDW	Caldwell - Essex County	NJ	Tower	
FRG	Farmingdale - Republic	NY	Tower	
ISP	Islip - Long Island MacArthur	NY	Tower	FSS
LDJ	Linden	NJ	Unicom	
MMU	Morristown - Municipal	NJ	Tower	
SWF	Newburgh - Stewart	NY	Tower	
N24	Spring Valley - Ramapo	NY	Unicom	
TEB	Teterboro	NJ	Tower	FSS
TTN	Trenton	NJ	Tower	
N87	Trenton-Robbinsville	NJ	Unicom	
HPN	White Plains - Westchester Cy	NY	Tower	

VFR Airports

N58	Hanover	NJ	Unicom
FLU	New York - Flushing	NY	Unicom
10N	Wallkill - Kobelt	NY	Unicom

Military Airfields

	McGuire AFB	NJ	Tower

OHIO

TCA: Cleveland (Group II) **FSS: Cleveland**

Main Airline Airport

CLE	Cleveland - Hopkins Intl	Tower	FSS

Other IFR Airports

BKL	Cleveland - Burke Lakefront	Tower
CGF	Cleveland - Cuyahoga County	Unicom
1G1	Elyria	Unicom
22G	Elyria - Lorain Cty. Regional	Unicom
LNN	Willoughby - Lost Nation	Tower

VFR Airports

4D6	Chardon		Unicom
4G8	Columbia Station		Unicom
7D6	Freedom - Liberty		Unicom
1G6	Strongsville		Unicom
3G3	Wadsworth - Municipal		Unicom
15G	Wadsworth - Weltzien		Unicom

Military Airfields

None

PENNSYLVANIA

TCA: Philadelphia (Group II) **FSS: Philadelphia**

Main Airline Airport

PHL	Philadelphia Intl	PA	Tower	

Other IFR Airports

N83	Bridgeport	NJ	Unicom	
N10	Collegeville - Perkiomen Val.	PA	Unicom	
N54	Langhorne - Buehl	PA	Unicom	
7MY	Mount Holly - Burlington Cty	NJ	Unicom	
PNE	Philadelphia - North	PA	Tower	FSS
N67	Philadelphia - Wings	PA	Unicom	
N46	Pottstown - Limerick	PA	Unicom	
N47	Pottstown - Municipal	PA	Unicom	
N34	Prospectville - Turner	PA	Unicom	
TTN	Trenton - Mercer County	NJ	Tower	
N87	Trenton - Robbinsville	NJ	Unicom	

VFR Airports

N99	West Chester - Brandywine	PA	Unicom	

Military Airfields

	McGuire AFB	NJ	Tower	
	Warminster NAS	PA	Tower	
	Willow Grove NAS	PA	Tower	

TCA: Pittsburgh (Group II) **FSS: Pittsburgh**

Main Airline Airport

PIT	Pittsburgh - Greater Intl.		Tower

Other IFR Airports

G01	Beaver Falls - Beaver County		Tower
BTP	Butler - County		Unicom

G08	Monongahela - Rostraver	Unicom	
AGC	Pittsburgh - Allegheny County	Tower	FSS
8G4	Pittsburgh - Campbell	Unicom	

VFR Airports

3G9	Butler - Farm Show-Roe	Unicom	
5G8	Jeannette - Pittsburgh/Boquet	Unicom	
4G0	Pittsburgh-Monroeville	Unicom	

Military Airfields

None

TEXAS

TCA: Dallas/Fort Worth (Group I) FSS: Dallas Fort Worth

Main Airline Airports

DFW	Dallas/Fort Worth - Regional	Tower	
DAL	Dallas - Love	Tower	FSS

Other IFR Airports

F54	Arlington - Municipal	Unicom	
F18	Cleburne - Municipal	Unicom	
ADS	Dallas - Addison	Tower	
RBD	Dallas - Redbird	Tower	
FTW	Fort Worth - Meacham	Tower	FSS
F72	Fort Worth - Oak Grove	Unicom	
F26	Plano - Dallas North	Unicom	
F46	Rockwall - Municipal	Unicom	
TRL	Terrell - Municipal	Unicom	
WEA	Weatherford - Parker County	Unicom	

VFR Airports

F09	Dallas - Air Park	Unicom	
1F7	Dallas - Doan East	Unicom	
F66	Desoto - Carroll	Unicom	
3F0	Fort Worth - Blue Mound	Unicom	
F64	Fort Worth - Flying Oaks	Unicom	
F71	Fort Worth - Luck	Unicom	
F70	Fort Worth - Mangham	Unicom	
F04	Fort Worth - Saginaw	Unicom	
9F9	Fort Worth - Sycamore	Unicom	
F67	Grand Prairie - Municipal	Unicom	
30F	Lake Dallas - Lakeview	Unicom	

20TX	Terrell - Phillips Ranch	Unicom	
71F	Terrell - Wallace	Unicom	
Military Airfields			
	Carswell AFB	Tower	
	Dallas NAS	Tower	
	Grand Prairie AANG	Tower	

TCA: Houston (Group II) **FSS: Houston**

Main Airline Airports			
IAH	Houston - Intercontinental	Tower	
HOU	Houston - Hobby	Tower	FSS
Other IFR Airports			
54TX	Baytown - RWJ	Unicom	
6R3	Cleveland - Municipal	Unicom	
AAP	Houston - Andrau	Unicom	
AXH	Houston - Arcola	Unicom	
HPY	Houston - Baytown	Unicom	
T02	Houston - Clover	Unicom	
SPX	Houston - Gulf	Unicom	
DWH	Houston - Hooks	Tower	
SGR	Houston - Hull	Unicom	
T86	Houston - Lakeside	Unicom	
T41	La Porte - Municipal	Unicom	
VFR Airports			
6R5	Alvin	Unicom	
1R9	Conroe - Cut and Shoot	Unicom	
T03	Genoa	Unicom	
39R	Houston - Beaman	Unicom	
T17	Houston - Weiser	Unicom	
T29	Pearland	Unicom	
TX10	Pearland - Skyway Manor	Unicom	
2XS3	Tomball - May	Unicom	
Military Airfields			
	Ellington AFB	Tower	

WASHINGTON

TCA: Seattle (Group II) **FSS: Seattle**

Main Airline Airport

SEA	Seattle - Tacoma Intl	Tower	

Other IFR Airports

PWT	Bremerton - Kitsap	Unicom	
RNT	Renton - Municipal	Tower	
BFI	Seattle - Boeing/King Country	Tower	FSS
S44	Spanaway	Unicom	
TIW	Tacoma Industrial	Tower	

VFR Airports

S50	Auburn	Unicom
BVU	Bellevue	Unicom
S36	Kent - Crest	Unicom
0S8	Port Orchard	Unicom
1S0	Puyallup - Pierce County	Unicom
S43	Snohomish - Harvey	Unicom

Military Airfields

	Gray AAF	Tower
	McChord AFB	Tower

Appendix B
Terminal Radar Service Areas (TRSAs)

State	Locations
AL	Birmingham, Huntsville, Mobile, Montgomery.
AK	Anchorage.
AZ	Phoenix, Tucson.
AR	Fort Smith, Little Rock.
CA	Burbank, Castle AFB (Merced), Monterey, Oakland, Ontario, Palm Springs, Sacramento, San Diego, Santa Ana.
CO	Colorado Springs.
CT	Windsor Locks (Bradley).
FL	Daytona Beach, Fort Lauderdale, Jacksonville, Melbourne, Orlando, Rensacola, Tallahassee, Tampa, West Palm Beach.
GA	Augusta, Columbus, Macon, Savannah.
HI	Kahului.
ID	Boise.
IL	Champaign, Moline, Peoria, Rockford, Springfield.
IN	Evansville, Fort Wayne, Indianapolis, South Bend.
IA	Cedar Rapids, Des Moines.
KS	Wichita.
KY	Cincinnati (Covington), Lexington, Louisville.

State	Locations
LA	Baton Rouge, Lafayette, Lake Charles, Monroe, Shreveport.
ME	Bangor, Portland.
MD	Baltimore.
MI	Flint, Grand Rapids, Kalamazoo, Lansing, Muskegon, Saginaw.
MN	Duluth, Rochester.
MS	Gulfport, Jackson.
MT	Billings, Great Falls.
NB	Lincoln, Omaha.
NV	Reno.
NJ	Atlantic City.
NM	Albuquerque.
NY	Albany, Binghamton, Buffalo, Islip, Rochester, Rome, Syracuse.
NC	Asheville, Charlotte, Fayetteville, Greensboro, Raleigh-Durham, Wilmington.
ND	Fargo.
OH	Akron-Canton, Columbus, Dayton, Toledo, Youngstown.
OK	Oklahoma City, Tulsa.
OR	Portland.
PA	Allentown, Erie, Harrisburg, Wilkes-Barre.
PR	San Juan.
RI	Providence.
SC	Charleston, Columbia, Greer, Shaw AFB (Sumter).
SD	Sioux Falls.
TN	Bristol, Chattanooga, Knoxville, Memphis, Nashville.
TX	Amarillo, Abilene, Austin, Beaumont, Corpus Christi, El Paso, Longview, Lubbock, Midland, San Antonio.
UT	Salt Lake City.
VT	Burlington.
VA	Dulles, Norfolk, Richmond, Roanoke.
WA	Spokane, Tacoma.
WV	Charleston, Huntington.
WI	Green Bay, Madison, Milwaukee.

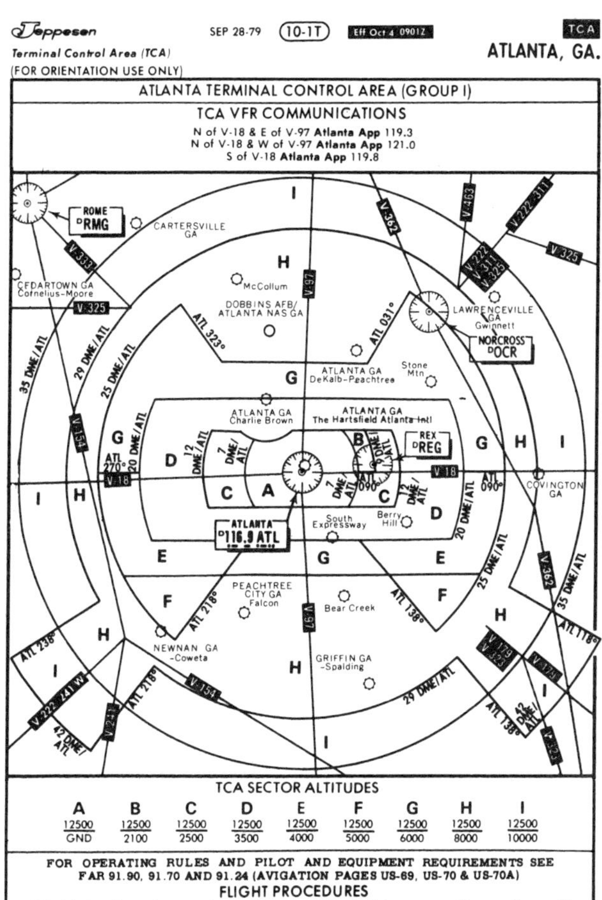

TCA SECTOR ALTITUDES

A	B	C	D	E	F	G	H	I
12500	12500	12500	12500	12500	12500	12500	12500	12500
GND	2100	2500	3500	4000	5000	6000	8000	10000

FOR OPERATING RULES AND PILOT AND EQUIPMENT REQUIREMENTS SEE FAR 91.90, 91.70 AND 91.24 (AVIGATION PAGES US-69, US-70 & US-70A)

FLIGHT PROCEDURES

IFR Flights-Aircraft operating within the TCA shall be operated in accordance with current IFR procedures. A clearance for a visual approach is not authorization for an aircraft to operate below the designated floors of the TCA.

VFR Flights-

a. Arriving aircraft should contact Atlanta App/Dep control on specified frequencies. Although arriving aircraft may be operated beneath the floor of the TCA on initial contact, communications should be established with Atlanta approach control for sequencing and spacing purposes.

b. Departing aircraft prior to taxiing are requested to advise the ground controller the intended altitude and route of flight to depart the TCA.

c. Aircraft not landing/departing The Hartsfield Atlanta Intl Airport may obtain an ATC clearance to transit the TCA when traffic conditions permit, provided the requirements of FAR 91 are met. Notwithstanding this, VFR transiting traffic is encouraged to the extent possible to fly beneath, above or around the TCA.

CHANGES: Horizontal and Vertical limits.

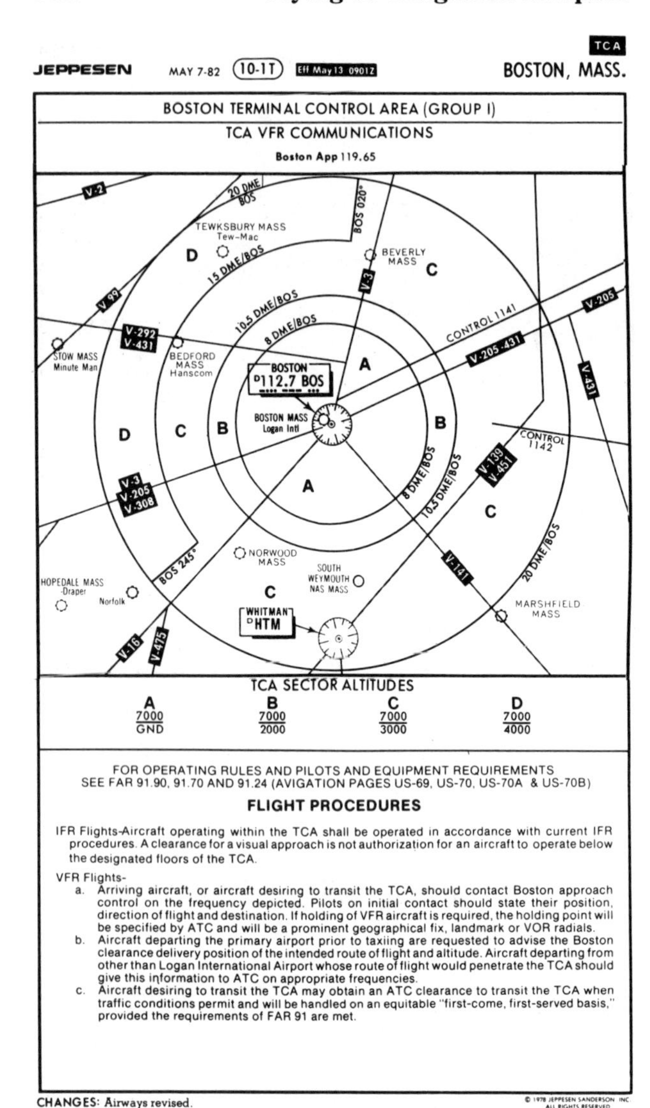

©1983 Jeppesen-Sanderson. Reduced for illustration only. Not for use in navigation.

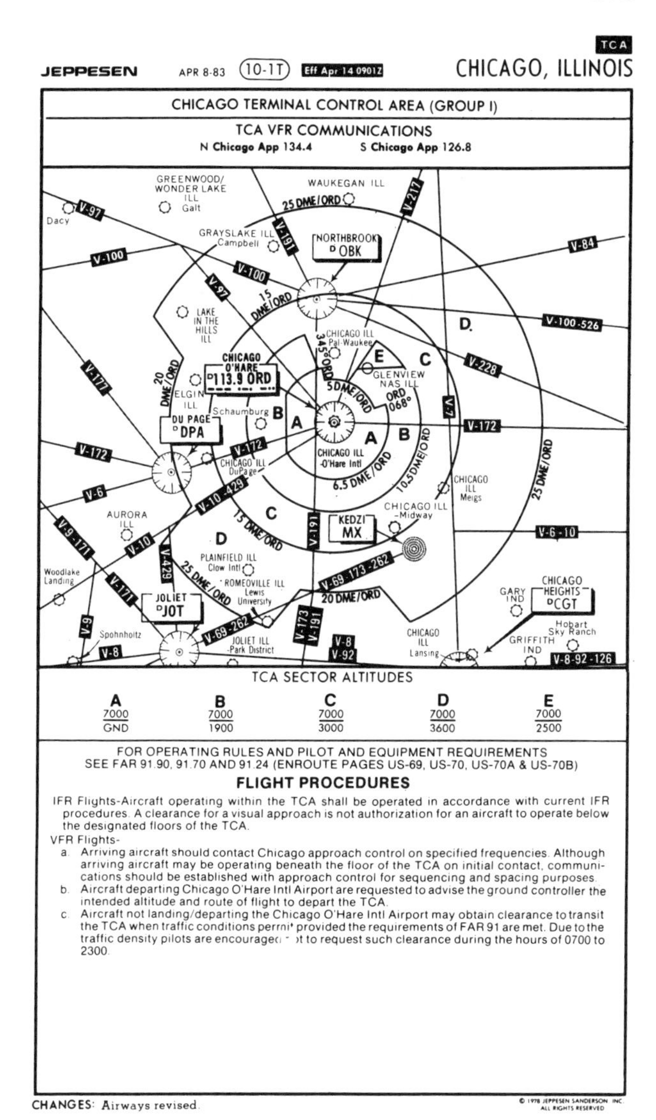

CHICAGO TERMINAL CONTROL AREA (GROUP I)

TCA VFR COMMUNICATIONS

N **Chicago App** 134.4 S **Chicago App** 126.8

TCA SECTOR ALTITUDES

A	B	C	D	E
7000	7000	7000	7000	7000
GND	1900	3000	3600	2500

FOR OPERATING RULES AND PILOT AND EQUIPMENT REQUIREMENTS SEE FAR 91.90, 91.70 AND 91.24 (ENROUTE PAGES US-69, US-70, US-70A & US-70B)

FLIGHT PROCEDURES

IFR Flights-Aircraft operating within the TCA shall be operated in accordance with current IFR procedures. A clearance for a visual approach is not authorization for an aircraft to operate below the designated floors of the TCA.

VFR Flights-

a. Arriving aircraft should contact Chicago approach control on specified frequencies. Although arriving aircraft may be operating beneath the floor of the TCA on initial contact, communications should be established with approach control for sequencing and spacing purposes.

b. Aircraft departing Chicago O'Hare Intl Airport are requested to advise the ground controller the intended altitude and route of flight to depart the TCA.

c. Aircraft not landing/departing the Chicago O'Hare Intl Airport may obtain clearance to transit the TCA when traffic conditions permi' provided the requirements of FAR 91 are met. Due to the traffic density pilots are encourage… …t to request such clearance during the hours of 0700 to 2300.

©1983 Jeppesen-Sanderson. Reduced for illustration only. Not for use in navigation.

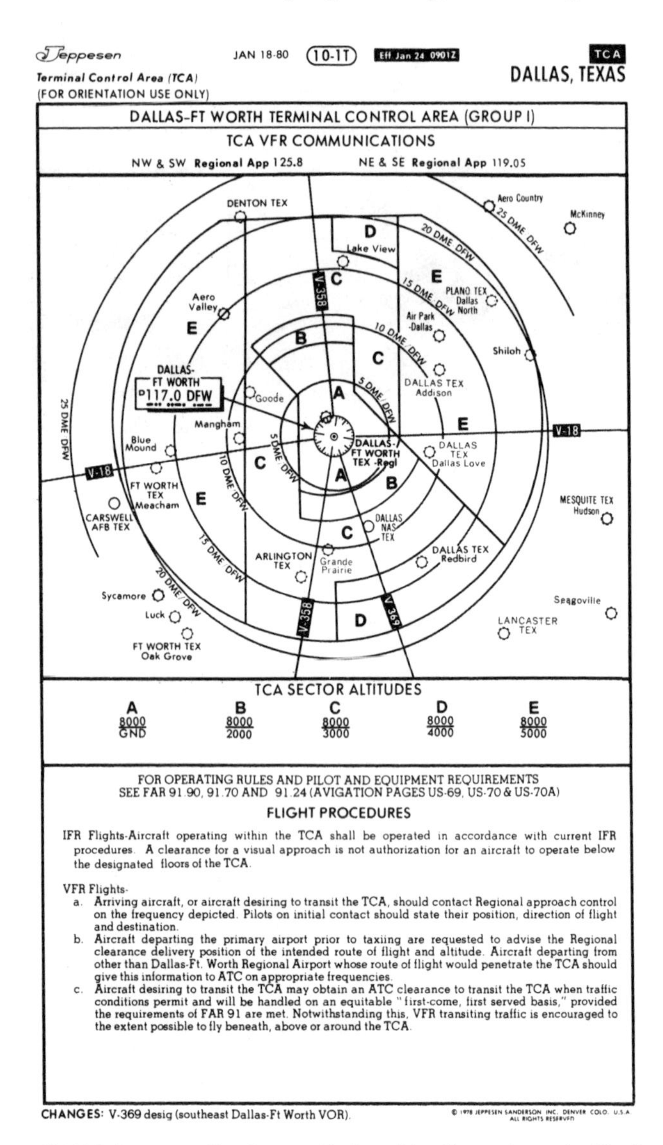

Jeppesen JAN 18-80 (10-1T) Eff Jan 24 0901Z TCA

Terminal Control Area (TCA)
(FOR ORIENTATION USE ONLY)

DALLAS, TEXAS

DALLAS-FT WORTH TERMINAL CONTROL AREA (GROUP I)

TCA VFR COMMUNICATIONS

NW & SW **Regional App** 125.8 NE & SE **Regional App** 119.05

TCA SECTOR ALTITUDES

A	B	C	D	E
8000	8000	8000	8000	8000
GND	2000	3000	4000	5000

FOR OPERATING RULES AND PILOT AND EQUIPMENT REQUIREMENTS
SEE FAR 91.90, 91.70 AND 91.24 (AVIGATION PAGES US-69, US-70 & US-70A)

FLIGHT PROCEDURES

IFR Flights-Aircraft operating within the TCA shall be operated in accordance with current IFR procedures. A clearance for a visual approach is not authorization for an aircraft to operate below the designated floors of the TCA.

VFR Flights-

a. Arriving aircraft, or aircraft desiring to transit the TCA, should contact Regional approach control on the frequency depicted. Pilots on initial contact should state their position, direction of flight and destination.

b. Aircraft departing the primary airport prior to taxiing are requested to advise the Regional clearance delivery position of the intended route of flight and altitude. Aircraft departing from other than Dallas-Ft. Worth Regional Airport whose route of flight would penetrate the TCA should give this information to ATC on appropriate frequencies.

c. Aircraft desiring to transit the TCA may obtain an ATC clearance to transit the TCA when traffic conditions permit and will be handled on an equitable "first-come, first served basis," provided the requirements of FAR 91 are met. Notwithstanding this, VFR transiting traffic is encouraged to the extent possible to fly beneath, above or around the TCA.

CHANGES: V-369 desig (southeast Dallas-Ft Worth VOR).

TCA

JEPPESEN JAN 30-81 (10-1T) LOS ANGELES, CALIF.

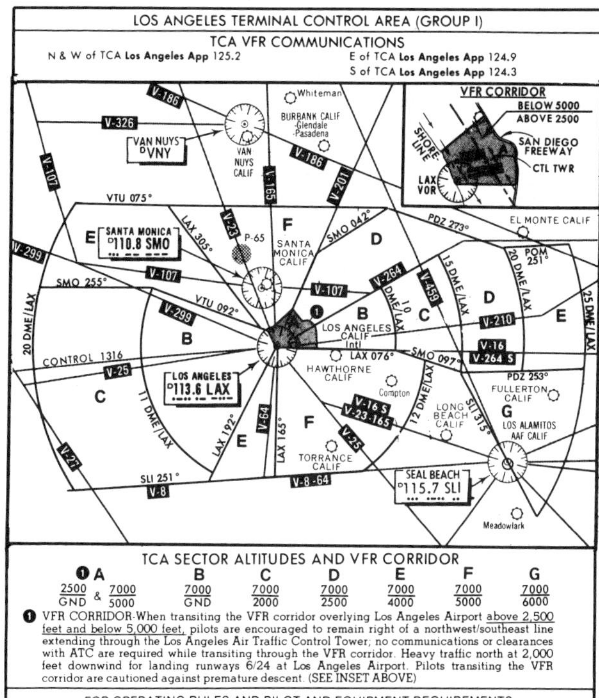
LOS ANGELES TERMINAL CONTROL AREA (GROUP I)

TCA VFR COMMUNICATIONS

N & W of TCA **Los Angeles App** 125.2

E of TCA **Los Angeles App** 124.9

S of TCA **Los Angeles App** 124.3

TCA SECTOR ALTITUDES AND VFR CORRIDOR

❶ A	B	C	D	E	F	G
2500/GND & 7000/5000	7000/GND	7000/2000	7000/2500	7000/4000	7000/5000	7000/6000

❶ VFR CORRIDOR-When transiting the VFR corridor overlying Los Angeles Airport above 2,500 feet and below 5,000 feet, pilots are encouraged to remain right of a northwest/southeast line extending through the Los Angeles Air Traffic Control Tower; no communications or clearances with ATC are required while transiting through the VFR corridor. Heavy traffic north at 2,000 feet downwind for landing runways 6/24 at Los Angeles Airport. Pilots transiting the VFR corridor are cautioned against premature descent. (SEE INSET ABOVE)

FOR OPERATING RULES AND PILOT AND EQUIPMENT REQUIREMENTS SEE FAR 91.90, 91.70 AND 91.24 (AVIGATION PAGES US-69, US-70, US-70A & US-70B)

FLIGHT PROCEDURES

IFR Flights-Aircraft operating within the TCA shall be operated in accordance with current IFR procedures. A clearance for a visual approach is not authorization for an aircraft to operate below the designated floors of the TCA.

VFR Flights-

a. Arriving aircraft or aircraft desiring to transit the TCA, should contact Los Angeles approach control on the frequency depicted. Pilots on initial contact should state their position, direction of flight and destination. If holding of VFR aircraft is required, the holding point will be specified by ATC and will be a prominent geographical fix, landmark or VOR radials.

b. Aircraft departing the primary airport prior to taxiing are requested to advise the Los Angeles clearance delivery position of the intended route of flight and altitude. Aircraft departing from other than Los Angeles International Airport whose route of flight would penetrate the TCA should give this information to ATC on appropriate frequencies.

c. Aircraft desiring to transit the TCA may obtain an ATC clearance to transit the TCA when traffic conditions permit and will be handled on an equitable "first-come, first-served basis," provided the requirements of FAR 91 are met. Notwithstanding this, VFR transiting traffic is encouraged to the extent possible to fly beneath, above or around the TCA or to transit north and south through the VFR corridor overlying Los Angeles Airport between 2,500 feet and 5,000 feet.

CHANGES: P-65 desig.

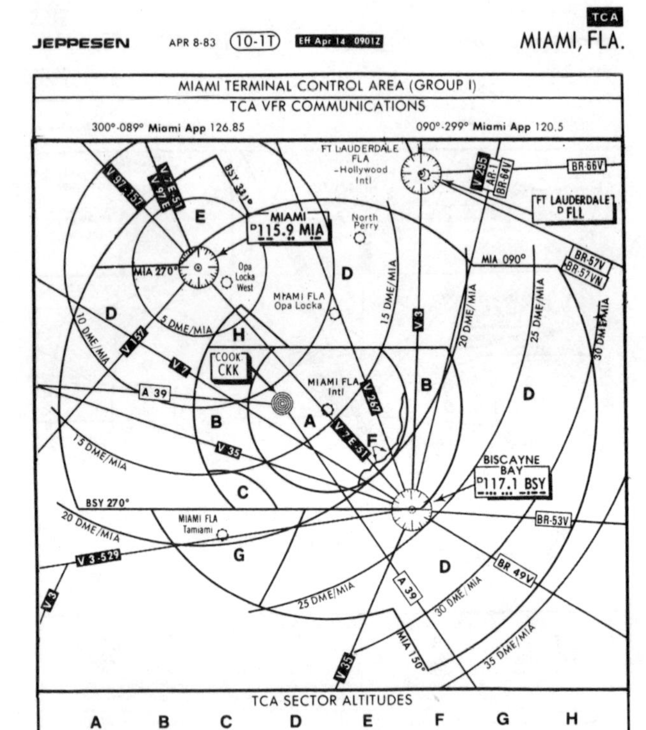

TCA SECTOR ALTITUDES

A	B	C	D	E	F	G	H
7000	7000	7000	7000	7000	7000	7000	7000
GND	1500	2000	3000	4000	1000	5000	2000

FOR OPERATING RULES AND PILOT AND EQUIPMENT REQUIREMENTS
SEE FAR 91.90, 91.70 AND 91.24 (ENROUTE PAGES US-69, US-70, US-70A & US-70B)

FLIGHT PROCEDURES

IFR Flights-Aircraft operating within the TCA shall be operated in accordance with current IFR procedures. A clearance for a visual approach is not authorization for an aircraft to operate below the designated floors of the TCA.

VFR Flights-

a. Arriving aircraft should contact Miami approach control on specified frequencies. Although arriving aircraft may be operating beneath the floor of the TCA on initial contact, communications should be established with approach control for sequencing and spacing purposes.

b. Aircraft departing the airport are requested to advise the ground controller the intended altitude and route of flight to depart the TCA.

c. Aircraft desiring to transit the TCA may obtain an ATC clearance to transit the TCA when traffic conditions permit and will be handled on an equitable "first-come, first-served" basis, provided the requirements of FAR 91 are met. Notwithstanding this, VFR transiting traffic is encouraged to the extent possible to fly beneath, above or around the TCA.

CHANGES: Airways revised.

Jeppesen JUN 27-80 (10-1T) TCA

Terminal Control Area (TCA)
(FOR ORIENTATION USE ONLY)

SAN FRANCISCO, CALIF.

SAN FRANCISCO TERMINAL CONTROL AREA (GROUP I)

TCA VFR COMMUNICATIONS

N **Bay App** 120.9 NE/E **Bay App** 135.4 SE **Bay App** 132.55
S **Bay App** 135.65 W **Bay App** 135.1

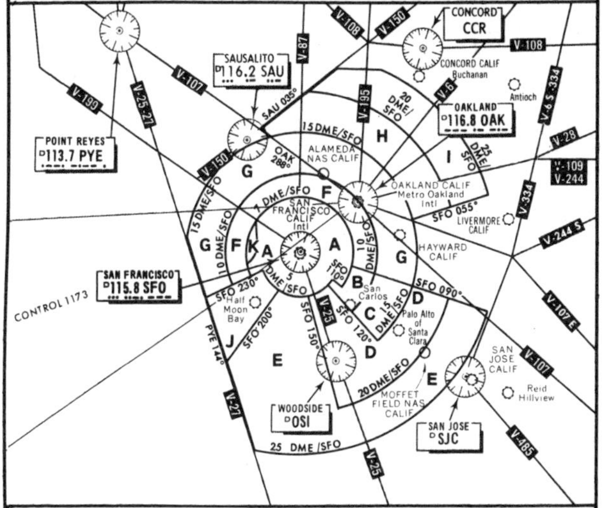

TCA SECTOR ALTITUDES

A	B	C	D	E	F	G	H	I	J	K
8000	8000	8000	8000	8000	8000	8000	8000	8000	8000	8000
GND	1500	2500	4000	6000	2100	3000	4500	6000	5000	1500

FOR OPERATING RULES AND PILOT AND EQUIPMENT REQUIREMENTS SEE FAR 91.90 91.70 AND 91.24 (AVIGATION PAGES US-69, US-70 & US-70A)

FLIGHT PROCEDURES

IFR Flights–Aircraft operating within the TCA shall be operated in accordance with current IFR procedures. A clearance for a visual approach is not authorization for an aircraft to operate below the designated floors of the TCA.

VFR Flights –

a. Arriving aircraft or aircraft desiring to transit the TCA, should contact Bay approach control on the frequency depicted. Pilots on initial contact should state their position, direction of flight and destination. If holding of VFR aircraft is required, the holding point will be specified by ATC and will be a prominent geographical fix, landmark or VOR radials.
b. Aircraft departing the primary airport prior to taxiing are requested to advise the San Francisco clearance delivery position of the intended route of flight and altitude. Aircraft departing from other than San Francisco International Airport whose route of flight would penetrate the TCA should give this information to ATC on appropriate frequencies.
c. Aircraft desiring to transit the TCA may obtain an ATC clearance to transit the TCA when traffic conditions permit and will be handled on an equitable "first-come, first-served basis," provided the requirements of FAR 91 are met.

CHANGES: VFR communications.

TCA

JEPPESEN JAN 30-81 (10-1T) **NEW YORK, N.Y.**

NEW YORK TERMINAL CONTROL AREA (GROUP I)

TCA VFR COMMUNICATIONS

LGA 071°-216° **New York App** 125.7 LGA 240°-328° **New York App** 125.5
LGA 216°-240° **New York App** 120.8 LGA 328°-071° **Westchester App** 120.55 126.4
2000' OR BELOW WITHIN 6.5 NM OF NEWARK INTL **Newark Twr** 127.85
2000' OR BELOW WITHIN 6 NM LA GUARDIA APT **La Guardia Twr** 126.05
2000' OR BELOW WITHIN 8 NM OF KENNEDY INTL **Kennedy Twr** 125.25

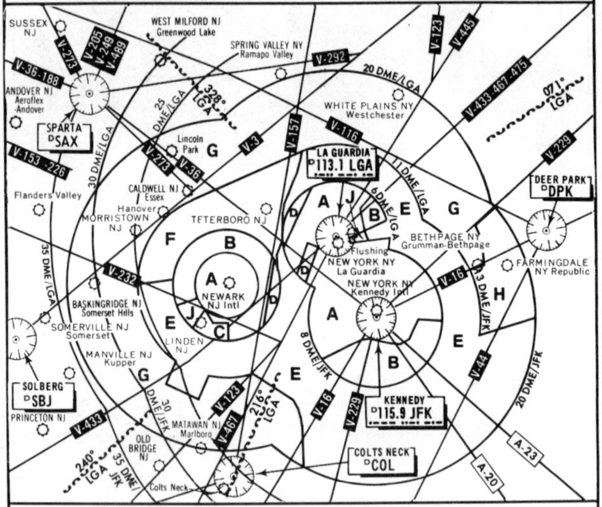

TCA SECTOR ALTITUDES

A	B	C	D	E	F	G	H	J
7000	7000	7000	7000	7000	7000	7000	7000	7000
GND	500	800	1100	1500	1800	3000	4000	1200

FOR OPERATING RULES AND PILOT AND EQUIPMENT REQUIREMENTS SEE FAR 91.90, 91.70 AND 91.24 (AVIGATION PAGES US-69, US-70, US-70A & US-70B)

FLIGHT PROCEDURES

IFR Flights–Aircraft operating within the TCA shall be operated in accordance with current IFR procedures. A clearance for a visual approach is not authorization for an aircraft to operate below the designated floors of the TCA.

VFR Flights –

a. Arriving aircraft, or aircraft desiring to transit the TCA should contact Approach Control on the frequency depicted for the sector of flight with reference to the LaGuardia VORDME. Pilots should state, on initial contact, their position, direction of flight and destination. If holding of VFR aircraft is required, the holding point will be specified by ATC and will be a prominent geographical fix, landmark or VOR radial/s.
b. Aircraft departing primary airports are requested to advise the appropriate clearance delivery position prior to taxiing of the intended route of flight and altitude. Aircraft departing from other than primary airports should give this information on appropriate ATC frequencies.
c. Aircraft desiring to transit the TCA will obtain clearance on an equitable "first-come, first served" basis, providing the requirements of FAR 91 are met.

CHANGES: Communications.

Jeppesen JUN 10-77 (10-1T) Eff Jun 16 0901Z TCA

Terminal Control Area (TCA)

(FOR ORIENTATION USE ONLY)

WASHINGTON, D.C. (VA.)

WASHINGTON TERMINAL CONTROL AREA (GROUP I)

TCA VFR COMMUNICATIONS

NE & SE **Washington App** 124.2 NW&SW **Washington App** 119.85

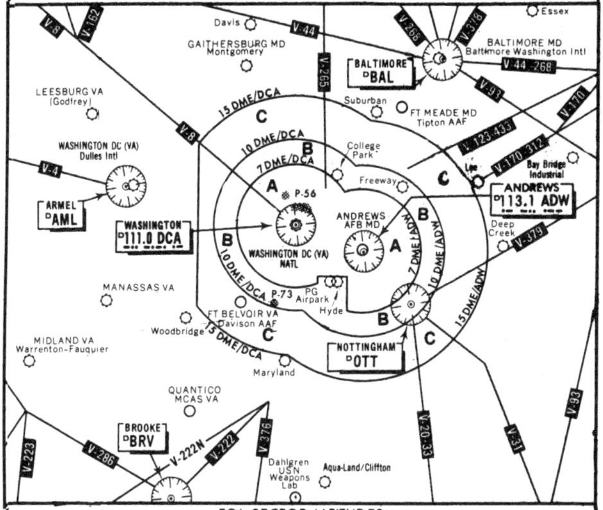

Davis
GAITHERSBURG MD Montgomery
V-8
V-162
V-44
V-265
V-268
V-378
Essex
BALTIMORE MD Baltimore Washington Intl
BALTIMORE ᴰBAL
V-44-268
V-93
V-170
LEESBURG VA (Godfrey)
15 DME/DCA
C
Suburban
FT MEADE MD Tipton AAF
V-123-433
V-170-312
WASHINGTON DC (VA) Dulles Intl
10 DME/DCA
B
7 DME/DCA
College Park
Lee
Bay Bridge Industrial
V-4
Freeway
ANDREWS ᴰ113.1 ADW
ARMEL ᴰAML
A
P-56
ANDREWS AFB MD
WASHINGTON ᴰ111.0 DCA
WASHINGTON DC (VA) NATL
7 DME/ADW
10 DME/ADW
15 DME/ADW
Deep Creek
V-379
MANASSAS VA
P-73
PG Airpark
Hyde
FT BELVOIR VA Davison AAF
Woodbridge
MIDLAND VA Warrenton-Fauquier
NOTTINGHAM ᴰOTT
Maryland
QUANTICO MCAS VA
BROOKE ᴰBRV
V-223
V-286
V-222N
V-222
V-376
Dahlgren USN Weapons Lab
Aqua-Land/Cliffton
V-20-33
V-31
V-93

TCA SECTOR ALTITUDES

A	B	C
7000	7000	7000
GND	1500	2500

FOR OPERATING RULES AND PILOT AND EQUIPMENT REQUIREMENTS SEE FAR 91.90, 91.70 AND 91.24 (AVIGATION PAGES US-69, US-70 & US-70A)

FLIGHT PROCEDURES

IFR Flights–Aircraft operating within the TCA shall be operated in accordance with current IFR procedures. A clearance for a visual approach is not authorization for an aircraft to operate below the designated floors of the TCA.

VFR Flights–

a. Arriving aircraft should contact Washington approach control on specified frequencies. Although arriving aircraft may be operated beneath the floor of the TCA on initial contact, communications should be established with Washington approach control for sequencing and spacing purposes.

b. Departing aircraft prior to taxiing are requested to advise Washington National clearance delivery or Andrews AFB ground control of the intended route of flight and altitude.

c. Aircraft not landing/departing the Washington National or Andrews AFB may obtain clearance to transit the TCA when traffic conditions permit, providing the requirements of FAR 91 are met. Notwithstanding this, VFR transiting traffic is encouraged to the extent possible to fly beneath, above or around the TCA.

CHANGES: V-31 redesig (SE of Nottingham).

Appendix C
Quiz Answers

Chapter 1 Quiz page 27.

1. a.
2. b.
3. c.
4. d.
5. b.
6. d.
7. b.
8. b.
9. d.
10. b.

Chapter 2 Quiz page 43.

1. d.
2. a.
3. a.
4. d.
5. a.
6. b.
7. a.
8. a.
9. d.
10. a.

Chapter 3 Quiz page 59.

1. d.
2. b.
3. b.
4. a.
5. c.

Chapter 4 Quiz page 87.

1. c.
2. b.
3. a.
4. b.
5. c.
6. a.
7. b.
8. c.
9. d.
10. a.

Chapter 5, Quiz page 98.

1. c.
2. b.

3. b.
4. a.
5. b.

Chapter 6 Quiz page 106.

1. b.
2. a.
3. d.
4. c.
5. b.

Chapter 7 Quiz page 123.

1. c.
2. b.
3. c.
4. c.
5. a.
6. b.
7. b.
8. a.
9. d.
10. b.

Chapter 8 Quiz page 141.

1. a.
2. a.
3. b.
4. d.
5. c.

Index